For Fred
with my very !

Rolf

Science AND Beyond

Toward Greater Sanity through Science, Philosophy, Art and Spirituality

Rolf Sattler

Suite 300 - 990 Fort St
Victoria, BC, V8V 3K2
Canada

www.friesenpress.com

Copyright © 2021 by Rolf Sattler
First Edition — 2021

All rights reserved.

No part of this publication may be reproduced in any form, or by any means, electronic or mechanical, including photocopying, recording, or any information browsing, storage, or retrieval system, without permission in writing from FriesenPress.

ISBN
978-1-03-910298-9 (Hardcover)
978-1-03-910297-2 (Paperback)
978-1-03-910299-6 (eBook)

1. Science, Research & Methodology.
2. Science, Philosophy & Social Aspects.
3. Science, Study & Teaching.

Distributed to the trade by The Ingram Book Company

"*Science and Beyond* is a powerful claim for integrating the limitations of science in our overall perception of reality, for recovering the fullness of life experience from the abstraction of scientific modelling, the unnamable from the named, wisdom and sanity from the infatuation with technological power. Such a return to balance and wholeness is crucially important not just for the survival and well-being of the human species, but also for the blossoming of life in all its forms."

— SHANTENA AUGUSTO SABBADINI, AUTHOR OF *TAO TE CHING: A GUIDE TO THE INTERPRETATION OF THE FOUNDATIONAL BOOK OF TAOISM* AND *PILGRIMAGES TO EMPTINESS: RETHINKING REALITY THROUGH QUANTUM PHYSICS*

Dedicated to

Alfred Korzybski (1879-1950), who understood science and could see beyond it,

Agnes Arber (1879-1960), whose investigations of plant form (plant morphology) led her to infinite issues beyond science,

Ravi Ravindra, professor emeritus of physics, philosophy, comparative religion, and a wise and compassionate human being,

Elisabet Sahtouris, evolution biologist and futurist, whose ecosophic vision encompasses evolution, ecology and economy, wisdom, and love,

and

His Holiness the 14th Dalai Lama, who embraces science, philosophy, and spirituality, wisdom and compassion.

Contents

ix	**Preface**
xiii	**Introduction**
1	**Chapter 1:** The Power of Science and Its Limitations
4	**Chapter 2:** Uncertainty and Beyond
10	**Chapter 3:** Replicability and Uniqueness
15	**Chapter 4:** Objectivity and Subjectivity
21	**Chapter 5:** Logic and the Indescribable
29	**Chapter 6:** Language and the Unnamable
38	**Chapter 7:** Empiricism and Just Sensing
48	**Chapter 8:** Mechanistic Materialist Science, Holistic Science, and Beyond
59	**Chapter 9:** Science, Society, and Culture
73	**Chapter 10:** Science and the Arts
77	**Chapter 11:** Science and Spirituality
84	**Conclusions**
89	**Appendix 1:** From Plant Morphology to Infinite Issues
96	**Appendix 2:** Health and Sanity of Body, Speech, and Mind
110	**Appendix 3:** The Human Condition and Its Transcendence
134	**References**
150	**About the Author**
152	**Index**

"Portal" (Networks) by Ulrich Panzer (2020)
Acrylic & Ink on Mylar, 19 x 17, 5 inches

Preface

Many books have been written on science. Why yet another one? Because **this book focuses on widespread misconceptions of science that can have grave and even disastrous consequences for our lives and society and that obscure the richness, profundity, infinity, and mystery beyond science.** These misconceptions have plagued us for a long time and have become incredibly destructive during the COVID-19 pandemic. Much suffering and mortality could have been avoided if the World Health Organization (WHO), chief medical officers, and governments had been sufficiently aware of these scientific fallacies. Thus, recognizing these fallacies can be a matter of health or sickness, life or death, which affects our existence and sanity in fundamental ways (see Chapter 9 together with Chapter 8 and the other chapters).

Contrary to many science books that have been written either by scientists or philosophers of science, this book is grounded in both a practical and philosophical understanding of science: I have carried out research in plant biology, especially plant morphology, for nearly forty years and have taught a course in the philosophy of biology, which included philosophy of science, for nearly thirty years. Thus, I have been deeply involved in the practical, theoretical, and philosophical investigation of science, which has led me to an inclusive and comprehensive view and experience of science (see Appendix 1).

Many books on science have been written by scientists who often have no or little knowledge of the philosophy of science. One can indeed carry out scientific research without such knowledge, but then the interpretation of the research may be limited and even misleading. To understand the full significance of science, we need an awareness of the often unrecognized philosophical presuppositions of science. Many philosophers of science have pointed out these presuppositions, but they usually lack the practical experience of scientific research. This experience that I have gained through my research adds an important dimension to the understanding of science.

I also taught courses on the history of biology, the relation of biology to society, and science and spirituality. Again, this book reflects these broader aspects of biology and science in general.

Through my teaching I have also learned from my students. And I have learned much from friends and other teachers, from scientists, philosophers, artists, and spiritual masters, including Alan Watts, Krishnamurti, Osho, Thich Nhat Hanh, the Dalai Lama, and others. I know that especially Osho remains controversial for several reasons. Nonetheless, I have learned much from him, especially about meditation. In addition to ancient meditations, he introduced meditations especially suited for our modern age. The Dalai Lama affirmed that "Osho is an enlightened master who works with all the possibilities to help humanity overcome a difficult phase in the development of consciousness."

From many wise men and women, I have learned that words and ideas, although they may convey great insights, remain limited, and therefore being dogmatically entrenched in any conceptual view limits us. I learned that ultimately, we have to transcend language, thought, and science that relies on thought and language. There are many ways of transcendence such as relaxation, meditation, and deep immersion in nature. Humour and laughter can also be liberating.

Although I emphasize transcendence, I want to stress that this book's title is not "Beyond Science," but "Science and Beyond." It shows how **a better understanding of science and its limitations will open the door to what is beyond these limitations that can lead to a richer, more fulfilled, happier, healthier, saner, and more peaceful life and society. If we cannot see and go beyond science, our lives and society will remain impoverished.**

Each chapter shows how, starting with science, we can transcend science to gain more inclusive vistas beyond science. I can see such vistas in "Portal," the painting by Ulrich Panzer (facing the preceding page). Thus, I can see this painting as an artistic portal to the topic of this book: *Science and Beyond*. Although a work of art, such as his painting, is always more than can be said about it, in the context of this book, I can see in "Portal" science, symbolized by the network of dots, leading to the mysterious source from which it arises. As highlighted on the front cover of this book: "The most beautiful thing we can experience is the mysterious. It is the source of all true art and science" (Albert Einstein, quoted by Ravindra 1991, p. 322).

I dedicate this book to **Alfred Korzybski** (1879-1950), author of *Science and Sanity* (1933), who understood science and could see beyond language and science that uses language. He referred to reality as the "unspeakable" that cannot be fully represented through language. Thus, he emphasized that language limits our understanding of reality in science and everyday life (Korzybski 1958).

I also dedicate this book to **Agnes Arber** (1879-1960) and **Ravi Ravindra**. Both worked as scientists and both went beyond science. Agnes Arber emphasized "The Manifold and the One" (1957) and Ravi Ravindra "Science and Spirit" (1991) and "Science and the Sacred" (2000).

I also dedicate this book to **Elisabet Sahtouris**. She developed an ecosophy that can lead us toward a wise society, a human way of life beyond the simplifications, separations, and misconceptions that are still so widespread in science and our society. Ecosophy comprises evolution, ecology and economy, finance, politics and government, science and spirituality, wisdom and love (Sahtouris 2014, 2018).

And I dedicate this book to **His Holiness, the 14th Dalai Lama,** who recognizes the importance of science and the wisdom and compassion beyond science. In his book *The Universe in a Single Atom* (2005b), he wrote: "I have argued for the need for and the possibility of a worldview grounded in science, yet one that does not deny the richness of human nature and the validity of modes of knowing other than the scientific…the union of wisdom and compassion"(see also Hayward and Varela 1992, Luisi 2009).

Acknowledgements – For valuable comments and suggestions I want to thank Joelle Hann, Madelyn Kent, Paul Mackenzie, Catherine Myers, Rolf Rutishauser and Gabriele Werle. To Somayeh Naghiloo I am grateful for drawing my attention to the *Speech of Birds*, also known as the *Conference of Birds*, by Attar of Nishapur (see at the end of Chapter 11). For the permission to reproduce illustrations I want to thank Somayeh Naghiloo, Ulrich Panzer, Rolf Rutishauser, Rory Skelly, Steve Stockdale, and Gabriele Werle.

I am grateful to the staff of FriesenPress, especially Scott Donovan for introducing me to FriesenPress and my publication specialist Brianne MacKinnon for guiding me through the intricacies of the production process.

Finally, I want to express my deep gratitude to mon amour Diane Ricard.

Chapter 5 is based on the first three chapters of *Healing Thinking and Being* published on my website https://www.beyondWilber.ca, and the three appendices are taken from the same website.

Introduction

The aim of this book is to create more awareness of:
1. widespread misconceptions of science,
2. grave and even disastrous consequences when individuals and society are misled by these misconceptions,
3. how these misconceptions obscure the profound limitations of science,
4. **the richness of life beyond the limitations of science that can lead to greater sanity, better health, more profound happiness and peace** (see also Appendix 2 and 3).

These issues affect nearly everybody, since science has become the dominant force in most parts of the world and our lives have become deeply enmeshed with science and technology. Therefore, these issues cannot be brushed away. They remain critical for our sanity, health and well-being, and for society. I felt compelled to write this book to convey how much misery could be alleviated through a better understanding of science, its limitations and beyond.

Science, Politics, and the Economy – It is probably generally recognized that science plays an important role in politics and the economy. Therefore, politicians and economists need to understand the power *and* the limitations of science. But do they see the limitations? If they did, they might see what lies beyond them: philosophy (as the love of wisdom), the arts, religion (as religious experience), ethics, and spirituality. I shall emphasize the importance of these and other areas beyond science throughout this book.

Science and the Arts – Although the importance of science is widely recognized, many would argue that the arts touch us more profoundly than science and therefore are more important than science. This is, however, much ignored by politicians and economists. Billions of dollars are allotted to scientific research and very little to the arts, which shows how dominant science has become and how much the arts have been neglected.

Science, Religion, Ethics, and Spirituality – Although there are still strongholds of antiscientific religious dogmatism in some parts of the world, many people try to reconcile religion with science or have turned away from religion. An increasing number of people say that they are spiritual but not religious (see, for example, Harris 2015). Although he is not against religion, the Dalai Lama (2011) emphasized the importance of ethics beyond religion. To a great extent, this turn beyond religion seems to stem from a rejection of the dogmatism that often pervades religions. However, religion, especially as religious experience, may transcend dogmatism, and spirituality is not always free of dogmatism. Non-dogmatic spirituality seems restricted to a relatively small segment of society. The majority of people in secular societies seem to have an unrealistic faith in science, especially mechanistic materialist mainstream science. Thus, more power is attributed to science than it has, which leads to delusion, a form of insanity.

To a great extent, this book is about sanity, how we can attain greater sanity through science, philosophy, art, and spirituality. To me, **sanity, in its deepest sense, means understanding and being in tune with life and reality.** Thus, insanity is not restricted to medical and psychiatric mental illness but includes any misconception or delusion about reality. When I refer to insanity, I mean it in this general sense. If you object to this usage of the term "insanity," you may replace it with "delusion" or "misconception." After all, one of the aims of this book is to point out widespread misconceptions of science. These misconceptions distort our view of reality at a very deep level and thus derange us profoundly. I refer to this derangement as insanity in its general and deepest sense to draw attention to the gravity of the situation.

How does sanity relate to **health**? Balance and wholeness appear saner than imbalance and fragmentation. Therefore, if, as in Chinese medicine, health is seen as balance and wholeness, it appears related to sanity (Appendix 2). Balance and wholeness require that, besides objectivity, we also value personal insight, subjectivity, uniqueness, and the body, in addition to the mind that figures so prominently in our cerebral Western culture. And we need a balance between the logical mind and the intuitive mind.

How does sanity lead to **happiness and peace**? Most people are happy when they obtain what they desire, especially when they can find their "other half" and then believe that their "other half" completes them and therefore makes them happy. But although a great lover, partner, or soulmate appears precious and among the greatest gifts one can obtain in this world, no other human being can complete us. Ultimate completion occurs in **oneness (nondualiy) with the universe, which leads to profound happiness and peace "independent of conditions"** (Shinzen Young 2016, p. 41). Therefore, throughout this book, I draw attention to our oneness (nonduality) with the universe, that "you are the universe" (Chopra and Kafatos 2017; more on this, especially in Chapter 11 and Appendix 3).

Painting by René Magritte that contrasts the common view of the separate and limited self with the unlimited Self that is one with its environment, which extends into the whole universe.

The complexity of the subject of this book may be a bit heavy to some readers. To compensate, I have tried to infuse some **lightness through a few jokes** (in smaller print so that the very serious reader who considers them inappropriate may skip them). Some of the jokes are taken from Osho's book *Take It Really Seriously. A Revolutionary Insight into Jokes* (1998), a collection of jokes from many sources. Osho explained that the purpose of a joke is not the joke itself but that it leads to laughter that transcends the thinking mind, if only for a moment.

Each chapter includes an invitation for a brief **contemplation or meditation.** Although these contemplations and meditations may appear too simple or even trivial, they may convey experientially the freedom beyond science, language, and thought and may give you a glimpse of oneness and nonduality if practiced with total dedication without intellectual preconceptions. I chose simple meditations to emphasize that even simple everyday life experiences can lead us beyond space and time into the infinite mystery. If you relax into these meditations, you may see the extraordinary in what often appears ordinary; you may see spirit in matter, the divine in the mundane, nirvana in samsara. If thoughts arise, accept them like clouds floating by in a clear sky. They, too, are part of the oneness and nonduality of existence. Oneness and nonduality do not exclude language, thought, and science: they include and transcend them (see, for example, Enza Vita. 2017. *Instant Presence: Allow Natural Meditation to Happen*).

Someone said the goal of meditation is to end suffering and to find the deepest home you have ever known. May the meditations in these chapters relieve or diminish your suffering and lead you toward your deepest home. Ultimately, everything in everyday life can become a meditation or a new way of being.

Each chapter ends with **conclusions and precious quotations** from great scientists, philosophers, artists, and spiritual masters.

An **extensive bibliography with many links** includes precious references, many of which are less known or forgotten, although they contain highly important and relevant insights and information. I consider this bibliography a precious part of the book that may allow the reader to pursue various topics in greater depth.

CONCLUSIONS

Science has become the dominant force in most parts of the world. But widespread misconceptions that may have grave or even disastrous consequences persist in the general population and even among many scientists. One aim of this book is to clear up these misconceptions.

Science has much to offer, but if we are unaware of its limitations and cannot see beyond them, our life remains impoverished. Similarly, a society that cannot see beyond science remains crippled. Thus, another aim of this book is to open up the way beyond science that may lead toward greater sanity, better health, more profound happiness, peace, and a wiser society.

As I shall show in the following chapters, science can reveal only aspects of the Truth, of Reality. If we believe that it reveals the Truth (that which is), we cannot see other aspects of the Truth. Mistaking an aspect of the Truth for the Truth involves delusion, a form of insanity. How to avoid this delusion, I shall point out throughout this book.

"Science isn't the answer to everything." (de Waal).

"Something would be needed beyond science." (David Bohm)
(Albert Einstein called David Bohm his "spiritual son"
and the Dalai Lama his "science guru")

Chapter 1
The Power of Science and Its Limitations

It is widely believed that science can be defined by its method, which is therefore referred to as **the scientific method**. This method is considered powerful. However, as Feyerabend has shown, there is no scientific method that characterizes all science and has been followed by all scientists (see Chapter 9). **Scientists arrive at testable laws, theories and models by a variety of methods. Based on these laws, theories and models, science has explanatory and predictive power**, which means that it can provide explanations and predictions. For example, it can explain why the sun rises, and it can predict at what time it will rise, or science can explain why people may get sick with an infectious disease, and it can predict that people who are in contact with a person who suffers from such a disease might also become sick. Furthermore, **science has also much power because of its technological applications**.

Explanation – As the above examples show, explanations may be deterministic or probabilistic, which means that they are derived from deterministic or probabilistic laws. Most laws and therefore most explanations are probabilistic. There are also explanations that are not derived from laws. Furthermore, explanation may be defined in different ways. According to a rather broad definition, explanation simply means analyzing or relating, or explanation is considered a synonym of clarification or understanding (see, for example, Sattler 1986, p. 54). Regardless of which definition we have in mind, **the power of scientific explanations and understanding appears limited**, as I shall point out throughout this book. **Not recognizing this limitation leads to delusions about reality, a form of insanity.**

Prediction – Many scientists and philosophers of science consider prediction more powerful than explanation. If a law or theory produces correct predictions, the law or theory is confirmed. Deterministic laws (if A, then B) allow deterministic predictions, which means accurate and reliable predictions, whereas probabilistic laws (if A, then probably B) allow only probabilistic predictions. It seems that the majority of laws and predictions, especially in the life sciences, are only probabilistic. Thus, for example, we cannot predict accurately whether a smoker will get lung cancer. We can only say that there is a probability that this will happen. **What actually happens in specific cases seems beyond the reach of science.**

Chaos Theory - The development of Chaos Theory in the second half of the twentieth century has also underlined the limited accuracy of predictions. Although based on deterministic equations, Chaos Theory does not provide accurate predictions for dynamical systems that are highly sensitive to initial conditions. Popularly, this is known as the **butterfly effect**, which means that "a small change in one state of a deterministic nonlinear system can result in large differences in a later state …A metaphor for this behaviour is that a butterfly flapping its wings in China can cause a hurricane in Texas" (Chaos Theory in Wikipedia). This explains, for example, why accurate and reliable predictions of the weather seem impossible. However, how a chaotic dynamical system develops can be indicated by a mathematical description called an attractor. An **attractor** is a set of numerical values that specifies a region toward which a system tends to evolve. This region seemingly "pulls" the system toward it. If the attractor has a fractal structure, it is called a **strange attractor** (a fractal structure is self-similar on all scales).

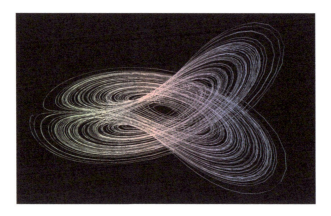

Image of a strange attractor

A CONTEMPLATION

With eyes open or closed, ask yourself: What is here right now if there is no problem to solve? You may find that if there is no problem to solve, you don't need your thinking mind, and thus you may transcend the thinking mind and enter universal spaciousness beyond language, thought, and science.

Simultaneously, you also transcend any references to the past or future that so often occupy our thinking mind more than necessary. Although the thinking mind is important, it is not necessary all the time. Many people cannot relax and have difficulties falling asleep because they cannot turn off their thinking mind when it is not needed (this contemplation is from Kelly 2015, pp. 28-29).

CONCLUSIONS

The desire of power pervades society in many ways. To a great extent, science and technology fulfill this desire of power that is rooted in our animal ancestry and is reinforced in Genesis of the Bible where we are told by God, "replenish the earth, and subdue it: and have dominion over …every living thing that moveth upon the earth" (see Appendix 3). However, the power of science, the accuracy and reliability of explanations and predictions have limitations. To avoid delusion, we have to be aware of these limitations. We have to realize that because of these limitations our personal decisions cannot be solely and reliably guided by science. Hence, in addition to science, we have to rely on **personal in-sight beyond science**. Needless to say, personal insight may also be limited, but as we learn to look deeply into ourselves in communion with nature, it may go beyond science "as if to be born from the womb of mystery" (Wolfe 2014, p. 40).

> "It is through the regular exposure to the contemplative experience that the faculties of intuition, insight, and realization are awakened, enabling us to appreciate the mystery of oneness" (George W. Wolfe).

> "Why stay in prison when the door is wide open?" (Rumi).

Chapter 2
Uncertainty and Beyond

NO PROOF IN SCIENCE

It is often assumed or even taken for granted that science can provide proof of its tenets, such as theories and laws, which means that scientific knowledge is proven knowledge. But **science cannot provide proof of its tenets. Science provides only evidence**. Sometimes the evidence for a scientific theory may be very strong. But even in this case we cannot tell whether future observations or experiments will confirm or contradict the theory. The history of science provides many examples of scientific revolutions where a well-established theory had to be modified or replaced by another one in view of new facts that could not be accommodated by the "established" theory. Newtonian physics is one such example. Some of its most fundamental assumptions, such as the absoluteness of time and space, had to give way to Einstein's relativity. Nonetheless, Newtonian physics still works as a special case, but it can no longer be regarded as *the* truth of the physical universe. Similarly, we cannot know whether future observations and experiments will confirm Einstein's relativity theories. Therefore, scientific knowledge remains open-ended, unproven. It lacks certainty. As Jamie Hale put it: "Scientific knowledge is tentative, and the tentative nature of science is one of its strong points." And yet we can read so often that this or that has been scientifically proven. Even scientists make such statements. This misconception is also prevalent in advertisements: so often we can read for a product or procedure "scientifically proven," which boosts sales to an uneducated public that is deceived by the illusory claim of scientific proof. On the other hand, products and procedures are devalued by the label "not scientifically proven" (which would be the appropriate label for *all* products and procedures). For example, the conservative medical establishment insists that alternative medicines are "not scientifically proven" and assumes that its products and procedures are often "scientifically proven." As a result, many people may prefer to use conventional medicine with potentially harmful or even fatal side effects, when they might have been helped or cured by alternative medicines (see, for example, Edwards 2007).

> Teddy Bearson is making his first visit to Doctor Bones.
> "And whom did you consult about your illness before you came to me?" Bones enquires.
> "Only the druggist down at the corner," replies Teddy.
> "And what sort of ridiculous advice did that fool give you?" demands Bones.
> "He told me," replies Teddy innocently, "to see you." (Osho 1998, p. 343).

In conclusion, many scientists and laypersons have a strong tendency to consider a scientific theory or tenet as proven when they see strong evidence for it and when they find strong recurring evidence. But **because of the open-endedness of science, not even the strongest evidence counts as proof.** Before the discovery of black swans in Australia, Europeans saw only white swans and thus concluded that all swans are white. However, seeing white swans again and again and again was no proof that all swans are white. So, next time you think you have proof, remember black swans.

Picture of a black swan

Believing in proof has many negative consequences that we can see in our society and individuals:
1. It closes the door to further enquiry. Why look for alternatives to a theory or law if it has been proven? For example, many practitioners of conventional medicine appear so imprisoned in the tenets of conventional medicine that they will not look beyond them for alternative medical practices. They have closed the door to alternative medicine because they believe that conventional medicine is proven and alternative medicine unproven.
2. It may create intolerance and dogmatism. Again, the example of medicine illustrates this. Many adherents of conventional medicine appear dogmatic and intolerant toward alternative medicine and work hard to suppress it.
3. It may lead to aggression in the name of the so-called proven knowledge. For example, practitioners of alternative medicine have sometimes been pursued in court, fined, or even imprisoned (see, for example, Ali 2012).
4. Since the belief in proof is based on delusion, it may lead to a degree of insane behaviour. According to Korzybski (1958, 2010), insanity implies a delusion

about reality. Believing that scientific knowledge is proven knowledge is not founded in reality.

A joke about Albert Einstein underlines the tentative nature of science:
Student: Dr. Einstein, Aren't these the same questions as last year's physics final exam?
Dr. Einstein: Yes; But this year the answers are different.

NO DISPROOF IN SCIENCE

Disproof is often referred to as falsification by philosophers of science. Like proof, it appears unattainable for several reasons. It seems that most scientists are not interested in disproving the theory or theories they like and in which they have a considerable investment. It would feel like pulling the rug from under their feet. Thus, psychologically, falsification does not seem appealing to most scientists. But psychology is not all that counts in science. Many would argue that objectivity is more important. Thus, the question: Is it possible to falsify (refute) a scientific theory or tenet? Popper (1962) and other philosophers answered this question affirmatively. They argued that if we can find one or more than one fact that contradicts a theory, this means that the theory is falsified and therefore should be modified or abandoned in order to remove the contradiction. Since facts are very important in science, this view seems to make sense. However, it bestows an absolute on facts that appears questionable. First of all, facts may be erroneous. To abandon a theory because it is contradicted by an erroneous fact would be a grave mistake. Secondly, facts are not as hard as often envisaged. They also contain a theoretical aspect and therefore are said to be theory-laden, which renders them low-level hypotheses. Obviously, a theory or hypothesis to be tested cannot be refuted by another hypothesis, even if that hypothesis, as a low-level hypothesis, may be much less hypothetical. And therefore, falsification appears to be impossible. Facts do not seem to have the force or "hardness" to refute theories in a definite and final way. Thus, whenever we find a clash between a theory and a fact, we can no longer have an absolute assurance that the theory is faulty; it might be the fact that's faulty or both the fact *and* the theory. Einstein once said: "If the facts don't fit the theory, change the facts." But I think he did not always have such disregard for facts.

Facts are theory-laden because they are observed in terms of a theoretical framework. Thus, what counts as a fact is at least in part determined by this theoretical framework. Let's look at a simple example of a fact, the statement "this leaf is green." It seems difficult to imagine how this fact could be theory-laden. But the concept "leaf" is part of the classical theory of plant morphology, according to which the shoots of plants consist of two kinds of organs: stem and leaf (see Appendix 1). According to another theory, the partial-shoot theory of the leaf (Arber 1950), a leaf is a partial shoot, and therefore the fact would be that "this partial-shoot is green." Furthermore, the concept "green" is also theory-laden because it reflects a theory of colour, such as the theory developed by Newton or the very different theory upheld by Goethe, the German poet and holistic scientist (see Bortoft 1996). Newton and Goethe pursued very different ways of scientific investigation. Newton's approach was analytical and mechanical, whereas Goethe emphasized the importance of intuition that may lead to an immersion with the phenomena beyond the subject-object division. **Nietzsche concluded: "There are no facts, only interpretations"** (see Cox 1999, 2, 3, 1)

Facts are not only theory-laden; they may also be value-laden, which means that we project values into them. Take, for example, the factual statement, "This animal is aggressive." Here, we may have projected aggressive behaviour, a prominent value in our competitive and aggressive capitalist society. Maybe the animal is just playful, or partly playful and aggressive, or whatever?

Another example: The war metaphor is often projected into nature and society (Lakoff and Johnson 1980, p. 4). Thus, one can hear talk about the conquest of nature or war against infectious disease. This view is then also projected into statements such as "The coronavirus SARS-CoV-2 attacks human health." It seems debatable to what extent facts and the whole scientific enterprise are value-laden. However, Scales (1995) concluded that "the image of science as pure value-free inquiry must be abandoned; there is no understanding nature independently of values" (see also Scales 2002). Nietzsche had come to the same conclusion: "all sense perceptions are permeated with *value judgments*" (see Cox 1999, 2, 3, 1). Nietzsche was a forerunner of postmodernism, which emphasizes the relativity of facts. However, relativity does not mean that they are totally fictitious. Facts remain important.

> "The theory, hypothesis, framework, or background knowledge held by an investigator can strongly influence what is observed." (Norwood Russell Hanson).

BEYOND UNCERTAINTY

Uncertainty appears in our experience within space and time. As we transcend this experience within space and time, we also go beyond uncertainty. Although such transcendence may be unimaginable to many people, to some people, it may happen spontaneously; to others in deep meditation. It may also occur in near-death experiences. Eben Alexander (2018) experienced it in a coma. In any case, it may lead to profound happiness and peace, independent of conditions (see Chapter 11 on Science and Spirituality).

If science always operates within space-time, then the transcendence of space-time would imply the transcendence of science. However, in quantum physics, it has been hypothesized that space-time may arise from a deeper reality beyond space-time (see Siegfried 2019). Thus, "quantum mechanics transgresses space and time in a very deep sense" (Zeilinger 2016). Does this mean that science can lead us to a realm beyond space and time and hence beyond uncertainty in our personal experience? Not really, because it cannot be proven since nothing can be proven in science. Remember black swans! Furthermore, it would remain a third-person perspective that cannot reach the first-person experience beyond uncertainty. To attain the first-person experience, we have to become "**metahuman**" (Chopra 2019). "Metahuman" means realizing our infinite potential, which includes uncertainty and beyond, science and beyond.

Paradoxically, a deep acceptance of uncertainty may also lead beyond it. Such acceptance removes the fear and thus allows us to relax into it, which leads beyond (Watts 1951, Brach 2003).

A MEDITATION

Close your eyes, inhale and then exhale deeply. While exhaling, let go completely. Let go of everything. If you are preoccupied with uncertainty or other worries, let go of all of them. If you find it helpful, you may chant "Aaaah" while exhaling. Repeat this until, with your exhalation, you feel like you're dissolving into the oneness of the universe beyond space and time, language, thought, and science.

CONCLUSIONS

Proof of scientific tenets, such as theories and laws, is not possible. Hence, we should give up all the talk about theories or laws being "scientifically proven." Unfortunately, such talk still pervades the public sphere and is not uncommon among scientists. Consequently, our society and individuals reap the insanity that results from such delusion. Since we cannot have proof, we can rely only on evidence. Uncertainty prevails. Heisenberg founded the uncertainty principle in physics. Subsequently, Korzybski (1933/1958) extended the uncertainty principle to other sciences and all domains of life. And Prigogine (1997) proclaimed the end of certainty.

Uncertainty is often seen as negative and threatening. However, some of the greatest scientists have underlined the positive sides of uncertainty. For example, the famous physicist Richard Feynman wrote: "I can live with doubt and uncertainty and not knowing…In order to make progress, one must leave the door to the unknown ajar" (quoted by Lane 2012). Alan Watts (1951) emphasized the "wisdom of insecurity" that is rooted in the acceptance of uncertainty.

Disproof (falsification) of scientific tenets also appears impossible, although this is not quite as obvious as the impossibility of proof. Since facts contain a theoretical element and therefore appear theory-laden and maybe even value-laden, they cannot disprove (falsify) a theory, although they seem to be less theoretical than a theory.

Believing that science can provide proof or disproof leads to a degree of insanity because this belief is not founded in reality. Insanity means being deluded about reality.

Since science cannot provide proof or disproof, we have to go beyond science in our ultimate decisions and follow our in-sight in addition to the available scientific evidence. **Personal insight**, although it may be more or less erroneous, personally and culturally conditioned, also has the potential to go beyond the categories of space and time. Since uncertainty appears in space and time, going beyond these categories means transcending uncertainty. Paradoxically, a deep acceptance of uncertainty may also lead beyond it. Such acceptance removes the fear and thus allows us to relax into it, which may lead beyond.

> "Wade through streams of time into the sanctuary of Being,
> and there, wait for the arrow of insight to slay all you think is real"
> (George W. Wolfe).

Chapter 3
Replicability and Uniqueness

REPLICABILITY

The replicability of results, such as the replicability of experiments, is considered the gold standard of scientific research. Replication supports the results, but it does not constitute proof because one cannot know whether future replication will be possible.

Failure of replication is usually taken as an indication that the results are untenable. Goldacre (2012) and Harris (2017) point out how sloppy science leads to a lack of replication. However, a lack of replication need not always be due to sloppy scientific procedures. One has to keep in mind that each time an experiment is repeated, some aspects of the context have changed. These aspects might be responsible for the lack of replication. For example, when an experiment is repeated, the moon phase may be different, and this might explain the failure of replication. Replication seems especially difficult in the soft sciences: biology, medicine, psychology, and the social sciences. Some have even referred to a **replication crisis** in these sciences (see Replication Crisis in Wikipedia).

> "Hymie Goldberg goes to see Doctor Feelgood in a terrible state. "You must help me, doctor," pleads Hymie. "I can't remember anything for more than a few minutes. It is driving me crazy."
> "I see," says the shrink. "And how long have you had the problem?" Hymie pauses, then says thoughtfully, "What problem?" " (Osho 1998, p. 646).

It is important to be aware of the **uniqueness of each context in which experiments are performed**. One may keep some variables of the context constant, but others are beyond our control. For example, we cannot control the eruption of solar flares. Thus, **a completely controlled experiment is not possible**.

The observer or experimenter also partakes in the context of an experiment. And in each new experiment, the experimenter differs, even if it is the "same" person. During one experiment, this person may be in a good mood, and when he repeats the experiment, he may be in a bad mood. If the experiment could not be replicated, was that failure due to his change in mood? How can we know? So many other features of the context may also have changed.

To reduce the influence of the experimenter, **randomized, double-blind experiments** are used, especially in medicine, psychology, and parapsychology. In a double-blind experiment, neither the experimenter nor the persons whose response is investigated know who receives the specific treatment. For example, to test the effect of a drug against a placebo, neither the experimenter nor the tested persons know who receives the drug and who receives the placebo. In this way, belief and expectations do not influence the outcome of the experiment. Therefore, this format has become the gold standard for many tests, such as the testing of medications. However, it does not provide proof, and the results of such experiments are not necessarily replicable because usually the experiment is carried out with a relatively small group of participants. And even if hundreds or thousands of participants were included, a repetition of the same experiment with different participants may not give the same results. But even if the results *could* be replicated, they may not extrapolate to the general population.

In some areas, such as the pharmaceutical industry and conventional medicine, the double-blind format has been elevated to an ideology. It is considered true science, whereas single-blind or non-blinded investigations are often denigrated as inferior science or not scientific at all. However, in many sciences, such as physics, the double-blind protocol is rarely used (Sheldrake 2020). Besides the advantages of double-blind studies, there are also disadvantages. Gaille (2020) listed **seven advantages and nine disadvantages of double-blind studies**, and Büller et al. (2009) pointed out that in certain situations the double-blind format may interfere with the objectives of the investigation, and therefore non-blinded experiments may provide a more accurate assessment of the situation. There are also many situations in which it is not desirable or is impossible to carry out double-blind investigations. For example, an acupuncturist cannot be blinded if he wants to compare an authentic acupuncture treatment with a fake one. And for the patient of an energy healer, it is not important to know whether effective healing is the result of the healing power of the healer or a partial or total placebo effect. What matters most for the patient is to be healed.

Although the double-blind format prevents bias created by the belief and expectations of the experimenters and the persons they investigate, besides belief and expectations, the very existence of the experimenter and the persons being investigated may influence the experiment. Furthermore, the wider context may also affect the outcome of an experiment. I heard that a researcher, after many successful replications of an experiment with rats, could one day no longer replicate it. He was very puzzled, but then found out that the technician who looked after the rats had been replaced by a new person. In contrast to the previous technician, the new person loved the rats and caressed them when he took them out of the cages for the experiments. It seems that this change in care explained why the outcome of the experiments could no longer be repeated. Harris (2017) pointed out that the gender of the handler in experiments with mice may also influence their outcome. A woman handling the mice may affect the experiment differently from a man handling them.

Although double-blind studies have advantages, one has to keep in mind that they are based on a limited sample, and therefore replication with another sample may fail. In contrast, folk knowledge and folk medicine, including traditional and indigenous medicine, although not double-blinded, are based on a long tradition and thus have been tested over hundreds or thousands of years. As a result, they may be remarkably successful. But they may also fail. Such failures often make headlines, whereas the failures of modern conventional medicine seem to be considered normal. Thousands of people die each day in hospitals around the world. In the United States alone, more than 700 thousand people die in hospitals every year.

UNIQUENESS

Although scientists may recognize the uniqueness of each event or object, such as each flower or each human being, their principal interest is not in the uniqueness. They abstract from the uniqueness properties shared by a class of objects, such as a species or a class of events like respiration. In doing so, they may arrive at interesting generalizations, one of the main aims of science, but they lose the uniqueness of each individual object or event (Sattler 1986). "According to Whitehead, the process of abstraction, useful as it may be…, is ultimately "false," in the sense that it operates by noting the salient features of an object [or event] and ignoring all else, and therefore "abstraction is nothing else than omission of part of the truth" (Wilber 1999, p. 80).

Uniqueness transcends replication because, by its very nature, it cannot be replicated. Therefore, those who believe that replicability constitutes absolute truth are deluding themselves. Replicability can at best lead to partial or relative truth, which can be seen as an aspect of Truth (that which is), what some call absolute truth.

We observe replication not only in experimental science but also in many natural events such as the daily cycles, moon cycles and the recurrence of seasons. But each sunrise, each new moon, and each recurrent season has uniqueness. So, we can embrace both the replication (repetition) and the uniqueness.

Beyond science, we can appreciate, love, and celebrate each unique object or event, each unique plant, animal, or human being. Therefore, let us not forget the preciousness of uniqueness.

Uniqueness is not static but flows. In this flow, any unique event connects with everything else, ultimately, the whole of existence. Being immersed in this flow can create **profound happiness** (Csikszentmihalyi 2008). The Dalai Lama said, "the very purpose of life is to seek happiness," and "the very motion of life is towards happiness" (Dalai Lama and Cutler 2009).

A picture of three unique flowers on a unique background.
Photo courtesy of Prof. Rolf Rutishauser.

A MEDITATION

Be totally attentive to a flower, a candle, or whatever, and through this attentiveness become one with the object and thus transcend language, thought, and science.

CONCLUSIONS

Although the replication of results remains an ideal in science, it cannot always be achieved and cannot even always be expected because, for each repetition of an observation or experiment, the relevant context may have changed and thus may have rendered replication impossible. We cannot have a completely controlled experiment because we cannot control all variables in the context.

Uniqueness, such as the uniqueness of a plant, an animal, or a human being, is beyond the scope and aim of science because scientists abstract (select) from the uniqueness of objects or events those features that are shared among them. Thus, they lose part of reality, which includes uniqueness, and they lose the Truth if Truth is defined as that which is. Hence, believing that replication gives us the Truth (that which is) leads to a degree of insanity because this belief is not founded in reality that includes uniqueness, which by its very nature cannot be replicated. To me, insanity means being deluded about reality.

Beyond science, we can include, appreciate, love, and celebrate each unique object or event, each unique plant, animal, or human being. Uniqueness is not static but flows. In this flow, any unique event connects with everything else, ultimately, the whole of existence. Being immersed in this flow can create profound happiness and peace.

"Listen to your own heart – be yourself, be original, be unique" (Osho).

"A human being is a single being. Unique and unrepeatable" (Eileen Caddy).

"One of the most important gifts a parent can give a child is the gift of accepting that child's uniqueness" (Fred Rogers).

"Each of us is a unique strand in the intricate web of life" (Deepak Chopra).

"You are unique and yet one with the universe" (Osho)

Chapter 4
Objectivity and Subjectivity

For many people, objectivity means reality or truth, but since we don't know the truth (that which is), this concept of objectivity is not useful in science. Instead, it seems more appropriate to understand objectivity as inter-subjectivity: thus, what is widely shared and accepted between subjects is considered objective (Sattler 1986, p. 65). For example, sharing the perception that humans have two legs makes it an objective datum. In contrast, seeing auras is usually considered subjective because this perception is not shared by many people. However, if the majority of people would learn to see auras, auras would become objective data. Whereas some people, including myself, can see only a more or less undifferentiated aura, others may see different layers within the aura.

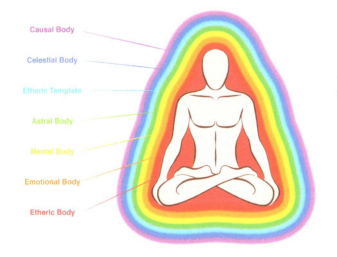

Diagrammatic representation of a sitting person with a multi-layered aura. Different names have been given to the layers.

Brennen (1988), in her book *Hands of Light*, distinguished seven or even nine layers. Since the ability to see auras depends on one's **state of consciousness**, the scientist's state of consciousness plays an important role in science and scientific progress. Kirchoff, referring to Rudolf Steiner, concluded: "A radical transformation of science can be brought about through a change in human consciousness… This will allow new features of the unrepresented to appear in full consciousness (Steiner 1912/1918). If this change is supported by a community of people who share the new state of consciousness, a new view of the world will emerge" (Kirchoff 1996, p. 23). What was subjective before will then become objective. Tart (1972) envisaged "**state-specific sciences**" of different scientific communities that share a specific state of consciousness. Each state represents a view from somewhere, whereas objectivity as reality or Truth would be a view from nowhere, hence no view at all (Callison and Young 2019). Since science cannot present a view from nowhere, it cannot reach objectivity as reality or Truth. However, the subjective experience and being of a meditator or enlightened being may transcend views from somewhere and lead to nowhere, which also means 'now here.' Lu-Tsu, the Chinese sage, said: "The land that is nowhere, that is the true home," which is also called "the still deeper secret of the secret" (Wilhelm 1962, p. 53) or "the secret of secrets" (Osho 1999, pp. 602, 619, 626).

Objectivity has also been defined as being free of bias. In the following I will show how **objectivity in this sense appears limited because of the expectancy effect, the experimenter and observer effect, selection bias, confirmation bias, consensus bias, reiteration bias, and publication bias.**

Expectancy Effect - It seems that most scientists are not free of expectations when they carry out research, and these expectations, which may be rooted in beliefs, influence not only the questions they ask but also how they go about finding answers to these questions. For example, a materialist who believes that all that exists is matter would only enquire into the material basis of diseases such as AIDS and find a material answer, whereas rare scientists, who can see beyond the materialist belief, can also ask how spirituality may influence the manifestation of AIDS, and they found that it plays a role in the progression of the disease (Church 2018, p. 189). For other examples that demonstrate the expectancy effect see Church (2018, pp. 189-197). He concluded: "If you expect something to happen, you're more likely to perceive it happening" (ibid, p. 189).

Observer and Experimenter Effect - An observation or experiment may be more or less influenced by the observer or experimenter because they interact with what is observed. This effect is well known in quantum physics, but it may also occur at the macro level. For example, depending on how you observe a person may lead to different reactions of that person. Different observers may also elicit different reactions. For example, a compassionate observer and an aggressive observer may find a different reaction. Animals and plants may also react differently to different observers and experimenters.

Wiseman and Schlitz (2019) carried out experiments about the effects of staring, whether people being stared at could feel it and turn their heads around. Wiseman had a negative attitude about these effects, and in his experiments failed to find any significant effects, whereas Schlitz had a positive attitude, and in her experiments found confirmation of the staring effects. Both Schlitz and Wiseman used the same experimental setup.

To counteract the experimenter effect, blind and double-blind experiments can be used, as I pointed out in the preceding chapter. According to Sheldrake, blinded methodologies are used mainly in the medical sciences and especially in parapsychology. In 2004, in the physical and biological sciences only up to 2.4% of all studies employed blinded methodologies (Sheldrake 2020). This percentage may have increased, but in many cases blinded methodologies may be undesirable, difficult or impossible, which leaves the vast majority of studies potentially biased by the experimenter effect. And even when in a double-blind experiment, neither the experimenter nor the participants in the experiment know who receives which treatment, in their unknowing they may nonetheless have an influence on the outcome of the experiment just through their existence.

Selection Bias involves the selection of data. It often leads to **confirmation bias** when we select data that support our preferred theory or worldview and neglect or ignore contradictory data. The selection may be conscious or subconscious. It happens when we become overly attached to our preferred views, which seems widespread among scientists and non-scientists. Although, in principle, conscious selection bias might be avoided, "a large number of psychological studies have shown that people [including scientists] respond to scientific or technical evidence in ways that justify their preexisting beliefs" (Teicholz 2015, p. 56). This leads to **belief perseverance**: people hang on to their beliefs. Most scientists do not try to refute their theories and world views. They do not follow Popper (1962), who thought that refutation (falsification) is the engine of scientific progress (see Chapter 2 and Sattler (1986) for a critique of Popper's view). Apart from some exceptions, most scientists seem to be happy when they can refute their opponents' views, but they don't like to refute their own ideas. Referring to T. C. Chamberlin, Teicholz (2015, p. 57) wrote that, "the moment you affix yourself to an idea an "intellectual child springs into existence," and it is difficult to remain neutral. The mind lingers "with pleasure" on the facts that support the theory." Nonetheless, there are probably some scientists who are not the victims of selection bias. However, even those who remain open-minded may not be aware of all the evidence that contradicts their views. The scientific literature has become too enormous to be known in its entirety by any one scientist. For example, who would be aware of a paper published long ago in an obscure French journal that challenges present-day research? I was lucky to find out about this paper through Gérard Cusset, a friend and colleague. Teicholz (2015) presents examples of relevant studies in nutrition science published long ago and have been forgotten or ignored by mainstream scientists.

Long ago, Goethe, the German poet and scientist, was well aware of confirmation bias when he wrote: "he [a scientist] will select from the data a few favourites that flatter him, he will manage to arrange the rest so that they will not appear to contradict him, and lastly he will complicate, obscure, and eliminate the hostile data" (quoted and translated by Mueller 1989, p. 224). But many people seem largely unaware of confirmation bias. Mechanistic materialist mainstream scientists select data that confirm mechanistic materialism and ignore, dismiss, or suppress contradictory data (see Chapter 8). During the COVID-19 pandemic, the World Health Organization and powerful medical officers, in collusion with the pharmaceutical industry and governments, discredited not only alternative medicine but also materialist scientists of mainstream science who disagreed with the official narrative (see Chapter 9). Goleman and Davidson (2017, Chapter 4) presented a critical review of confirmation bias and other kinds of bias in research on the benefits of meditation. For more examples of confirmation bias see Waller (2004).

Consensus Bias is a bias toward the majority view, involving the assumption that the majority must be right. There may be several reasons why scientists as well as laypersons may prefer to opt for the majority view, which may or may not be well supported by evidence. Following the herd may provide a sense of security. It may increase the chances of getting one's research published. It may increase the chances for funding of research. It may increase the chances of finding a position or being promoted. But it may be a hindrance to innovation and progress. An example of consensus bias is the widespread idea that consciousness originates in the brain (see Chapter 8).

Reiteration Bias, also called Illusory Truth Effect, is the tendency to believe what is often repeated. This effect is well known in political propaganda, election campaigns, advertising, and news media. Although science is commonly considered objective, scientists are not necessarily immune to reiteration bias. For example, repeating again and again that vaccination is the final solution to the COVID-19 pandemic became a widely accepted doctrine, although effective alternative treatments were already available during the crisis, but they were ignored and suppressed by the conservative medical establishment (see Chapter 9).

Publication Bias may occur for several reasons. It seems that scientists often publish only a portion of their data. For example, they may omit negative results because editors and publishers are more interested in positive results. Furthermore, editors, reviewers, and publishers may reject a paper or book because it is not to their liking for various reasons. In this way, poor research may be eliminated, but one wonders how much outstanding research has been rejected because it contradicted the editor's or reviewer's cherished beliefs.

Additional limitations of objectivity have been pointed out by Sheldrake (2020, Chapter 11: Illusions of Objectivity) and Hands (2017, Chapter 32: The Limitations of Science). Kirchoff (1996, p. 7) concluded: "The ideal of the objective scientist impartially weighting the data is a myth." The Nobel laureate Peter Medawar concurred: "**Innocent, unbiased observation is a myth**" (quoted by Hands 2017, p. 563). Church (2018, p. 189) concluded: "Belief permeates and shapes the entire field of science" (see also Reiss and Sprenger 2014). According to the filter model, our conditioned minds filter out aspects of reality.

Nonetheless, objectivity is upheld as an ideal that, however, cannot be reached if objectivity implies reality and the Truth. As I am pointing out throughout this book (see especially in Chapter 6 on language), **Truth appears unattainable in science. Science can give us at best partial truths, aspects of reality.** Confusing aspects of Truth with Truth can have grave and devastating consequences, one of them that one cannot recognize other aspects of Truth such as subjectivity, the arts, and spirituality. This confusion leads to an impoverished life and society (see, for example, Wallace 2000, Hawkins 2013, Glazier 2019). Of course, **subjective experience** may be more or less erroneous, personally and culturally conditioned, but it can also be the door to the most **profound wisdom** such as the wisdom of the Laozi, Buddha, Heraclitus, and many other sages of the East and West (see p. 81). They could see beyond the objectivity (as inter-subjectivity) of science, for whatever can be widely shared and accepted is not necessarily profound or true, although it may be an aspect of reality. Ignoring subjectivity narrows our window to reality because subjective experience may be a window to the world and reality, to oneness and nonduality of which we may become aware in meditation or spontaneously anywhere anytime.

Defining subjectivity just in terms of bias, as it is often done, overlooks **the deepest potential of subjectivity, namely that it may lead to insight, wisdom, happiness, and peace beyond science.** But unfortunately, bias is often involved in subjectivity and also in the so-called objectivity of science, as I pointed out above. It may be reduced or eliminated through a transformation of consciousness (see Chapter 11).

Finally, we can welcome and accept both subjectivity and objectivity (what is shared) and maybe transcend both (Ferrer 2017). In another sense, even in statistics one can go beyond subjectivity and objectivity (Gelman and Hennig 2017).

> "Hymie and Becky Goldberg are having a day in the country. Becky sees a lovely place under a tree next to a small pond and points it out to Hymie.
> "That's a beautiful spot for a picnic," she says.
> "It must be, dear," shrugs Hymie. "Fifty million mosquitos can't be wrong." " (Osho 1998, p. 89).

A MEDITATION

Look within and go beyond. There are many ways of doing this. One way is through inner smile: Close your eyes, relax, and then feel an inner smile in your heart or belly or your whole body. You may notice that with this smile you can feel connected to the infinity of existence beyond language, thought, and science.

In Daoism inner smile is used for healing. In holistic medicine healing is also called "wholing," which relates to wholeness and holiness.

CONCLUSIONS

For many people, objectivity means reality or Truth, but since we don't know the Truth (that which is), this concept of objectivity is not useful in science. Instead, it seems more appropriate to understand objectivity as inter-subjectivity: thus, what is widely shared and accepted between subjects is considered objective. What is widely shared reflects a collective state of consciousness. A change in the state of consciousness can lead to progress in science.

Objectivity has also been defined as being free of bias. However, objectivity in this sense appears limited because of the expectancy effect, the experimenter and observer effect, selection bias, confirmation bias, consensus bias, reiteration bias, and publication bias. Nonetheless, objectivity is upheld as an ideal that, however, cannot be reached if objectivity implies reality and the Truth. Truth appears unattainable in science. Science can give us at best partial truths, aspects of reality. If this is understood, then subjectivity as an alternative avenue to Truth cannot be ruled out. Of course, subjective experience may be more or less erroneous, personally and culturally conditioned, but it can also have an enormous richness and may be the door to the most profound wisdom such as the wisdom of the Laozi (Lao Tzu), the Buddha, and many other sages of the East and West who could see beyond objectivity (as inter-subjectivity) and science, for whatever can be widely shared and accepted is not necessarily profound or true, although it may be an aspect of reality. Defining subjectivity just in terms of bias, as it is often done, overlooks the deepest potential of subjectivity, namely that it may lead to insight, wisdom, happiness, and peace beyond science. But unfortunately, bias may be often involved in subjectivity and also in the so-called objectivity of science. It may be reduced or eliminated through a transformation of consciousness (see Chapter 11). Ultimately, we can encompass both subjectivity and objectivity (what is shared) and maybe transcend both.

> "Modern culture is constantly growing more objective. Its tissues grow more and more out of impersonal energies, and absorb less and less the subjective entirety of the individual" (Georg Simmel).

> "Subjectivity is incredibly valuable" (J. W. Shultz).

> "Indian thought is obsessed with subjectivity, Greek thought with objectivity" (Devdutt Pattanai). One can also have both without obsession.

> "Every physical process may be said to have objective and subjective features" (Niels Bohr).

Chapter 5
Logic and the Indescribable

Scientific knowledge is expressed through language that implies logic and mathematics, a form of language and logic. Except for quantum physics and a few other disciplines, **most sciences still use Aristotelian logic** and often take it for granted or even assume that Aristotelian logic is the only logic available. However, as I shall show below, other kinds of logic have existed for thousands of years. But Aristotle has had an enormous influence on Western culture and science. We have been conditioned by his logic from early childhood. Hence, we use it most of the time and may not even be aware of it. We use it in everyday life, in many sciences, in law, in politics, even in religion.

Aristotelian logic is based on what is often referred to as the laws of thought (see, for example, Edwards 1967; Arber 1964, p. 82).

These so-called laws of thought are:
1. **The law of identity: A is A.**
2. **The law of contradiction (also called the law of noncontradiction): A is not both A and not-A.**
3. **The law of the excluded middle: A is either B or not-B, or "everything is either A or not-A"** (Edwards 1967, Vol.4, p. 414).

Beyond the Laws of Thought - Although the law of identity seems unquestionable, it is not. For example, I am I, according to this law. But I am also the universe. Some mystics have indeed said: I am the universe. Such a statement is based on subjective experience, but it is also compatible with objective scientific evidence (see, e.g., Hollick 2006, Chopra and Kafatos 2017). Science has shown the interconnectedness of everything. Hence, I am connected with my environment, which extends into the whole universe. Being connected with the universe means that I am one with the universe, which is another way of saying that I am the universe. I am not just I.

Identity exists only in logic and mathematics. If you have a close look, you cannot find identity in the real world (see, for example, the Dalai Lama 2006). Yet too many people have created a **cult of identity**. And to a great extent, our society supports this cult. As a consequence, people become divided by their different "identities," which creates a basis for antagonism, conflict, and even war. But one could also argue that "identity" creates a sense of belonging as one "identifies" with a group, an ideology, or religion. Yet this kind of belonging creates false security, which is not based in reality. As

Heraclitus, the ancient Greek philosopher, wrote: Everything flows, everything changes. You cannot step into the same river twice, because next time the river has changed and you have changed. So, you are not even identical to yourself as time passes. How can you be identical to something else?

The second law of thought, the law of contradiction, can also be questioned. I am I (in the usual restrictive sense), but since I am also my environment or the universe, which is not-I (in the restrictive sense), I am I and not I, which contradicts the second law of thought, the law contradiction.

And finally, the third law of thought, the law of the excluded middle, is also very questionable. For example, according to this law, everything would be either black or white. But our world is not just black and white; it also contains many shades of grey.

Below I shall give more examples that illustrate the questionable status of the laws of thought. I will not conclude that these laws are useless. But I want to point out their limited usefulness. If they are taken for granted and used exclusively, as is often the case, they may become harmful to the individual, society, and the planet.

Logicians have known for a long time that the so-called laws of thought are limited "because no viable system of logic can be constructed in which the principles of identity, contradiction, and excluded middle would be the only axioms" (Edwards 1967, Vol. 4, p. 414). However, to a great extent, the general public and even many scientists have remained unaware of this limitation and its harmful consequences.

Yin-Yang Thinking - Alternatives to the so-called laws of thought have been available for millennia. In ancient China, the Daoists developed Yin-Yang thinking. The diagrams below illustrate the difference between Yin-Yang thinking and Aristotelian thinking based on the laws of thought.

The square to the left illustrates Aristotelian thinking based on the laws of thought, the Yin-Yang symbol to the right, Yin-Yang thinking.

If we call the black portion of the square A, then A is A (the law of identity), and A is not not-A, the white portion (the law of contradiction). And if we refer to the whole square, everything is either black or white (A or B, which is not-A).

In contrast, in the Yin-Yang symbol, Yang, the black area, includes Yin, the white dot, and Yin includes Yang, the black dot. Therefore, if we call the black half again A, A is A and not-A, and the same applies to the white half (which negates the laws of identity and contradiction). We also see that Yang gradually merges with Yin and vice versa, which indicates fuzziness: something may be Yang or Yin, A or B, to various degrees (which negates the law of the excluded middle).

The diagrams indicate why Aristotelian logic based on the so-called laws of thought can be harmful, whereas Yin-Yang thinking can be healthy and healing. The laws of thought cut reality into pieces, into opposites that are disconnected and may become antagonistic and destructive. They create wounds. Yin and Yang are also opposites, but they are connected. Through this connection, they can heal the split that has been created by the laws of thought. For example, if I say, "I am good and you are bad" and this is understood in terms of the laws of thought, as it is typical in our culture, I create a disconnection that may lead to antagonism, conflict, and harm. If, however, I look at this situation in terms of Yin-Yang, I understand that I am also bad and you are also good, and in this way, we are connected despite our opposition. And this connection heals the wound that has been created by the cut through the laws of thought. Hermann Hesse, in his Siddhartha, understood this situation very well when he wrote: "But the world itself, being in and around us, is never one-sided. Never is a man or a deed wholly Samsara or wholly Nirvana; never is a man wholly a saint or a sinner (Hesse 1951, p. 113).

Besides Yin-Yang thinking, Buddhist and Jain logic also transcend our common Aristotelian logic. Going beyond this limited and limiting logic can lead to greater sanity, better health, more profound happiness and peace (see Appendix 2).

Buddhist Logic – In contrast to Aristotelian logic, which is two-valued (either - or), Buddhist logic, as developed by Nagarjuna, is four-valued. In addition to either - or, it includes both/and as well as neither/nor. According to Nagarjuna, the Buddha first taught that the world is real. He next taught that it is unreal. To the more astute students, he taught that it is both real and not real. And to those who were furthest along the path, he taught that the world is neither real nor not real. Concerning good and bad (evil), we would then conclude that a person, organization, group, or nation is good, bad, both good and bad, and neither good nor bad. With regard to truth, any statement would have the values true, false (not true), both true and false, neither true nor false.

Buddhist logic can be liberating because it transcends not only the restrictive either/or of our common way of thinking, but even the both/and of the much more inclusive both/and logic. When we conclude that something is neither this nor that – neti neti as

Hindus (especially Advaita Vedantists) would say – we transcend thought and thinking altogether. We point to the unnamable, the mystery beyond the thinking mind (see also the following chapter).

Reaching beyond the thinking mind can be deeply healing. It removes the agitation of the thinking mind and delivers us into the calmness of infinite wisdom. But note that Buddhist logic does not exclude either/or and both/and. However, seen from the perspective of neither/nor, either/or appears less threatening and less harmful because its limitation is recognized.

> A philosopher tells a friend she's had a baby. The friend says "Congratulations, is it a boy or a girl?" The philosopher says "yes."

Jain Logic (or Jaina Logic), which was developed in Jainism, is seven-valued (Rankin 2010, p. 14; Diem-Lane 2020). Maybe even more than Buddhist logic, Jain logic recognizes the complexity of reality. Since no single proposition can capture this complexity, every proposition should be prefixed by the term "syad," which in the context of Jain logic means "in some ways" or "from one perspective." The following seven forms of "syad," seven perspectives, seven logical values have to be acknowledged:

1. "in some ways it is"
2. "in some ways it is not"
3. "in some ways it is and it is not"
4. "in some ways it is and it is indescribable"
5. "in some ways it is not and it is indescribable"
6. "in some ways it is, it is not and it is indescribable"
7. "in some ways it is indescribable"

This form of seven-valued logic avoids dogmatism, antagonism, and conflict. Since any one statement is not the full truth but only one of seven perspectives, it leaves room for the other six perspectives. Thus, even opposites are included. And since the indescribable is admitted, language, logic, and thinking are transcended: the unnamable, the mysterious, which is beyond the grasp of the thinking mind, is acknowledged. For this reason, Jain logic appears to be the most comprehensive logic, far beyond the scope of our common either/or logic that admits only two mutually exclusive values. Jain logic also includes these two values of our common logic, but they are no longer absolutes; they are only two perspectives among the other perspectives. Jain logic has been called "many-sided wisdom" (Rankin 2010). It entails perspectivism (see Sattler 1986, 2018).

Jain logic is particularly suited to heal the wounds that have been created by the harmful thinking of our common binary logic. Let us look at the simple statement "He is bad," whose opposite is "He is good (not bad)." According to Jain logic, the situation is not at all simple because it encompasses all of the following: "In some ways, he is bad." "In some ways, he is not bad." "In some ways, he is bad and not bad" (both/and logic). "In some ways, he is bad and indescribable." "In some ways, he is not bad and indescribable." "In some ways, he is bad and not bad and indescribable." "In some ways, he is indescribable."

Often, we can see only one perspective and believe that it is the full truth. We are convinced that he is bad, or we are convinced that he is good, and then we act according to such narrow-minded convictions in a way that can be harmful. Jain logic helps us to transcend such convictions, narrow-mindedness, and dogmatism so that we can see a much richer spectrum of reality. Action, based on this much more comprehensive view and understanding, can be more beneficial and peaceful.

Logic in Modern Science – As far as I know, Jain logic has not been fully incorporated into modern science. However, in quantum physics and some other holistic sciences, the two additional values of Buddhist logic (neither/nor and both/and) have been accepted. In quantum physics, it became obvious that Aristotelian either/or logic could not sufficiently deal with the observed phenomena. An electron could manifest both as a particle and wave depending on the experimental setup, that is, depending on the mode of observation. Therefore, the question of whether it is either a particle or a wave did no longer make sense. The either/or had to be dropped in favour of both/and. But it would be incorrect to conclude, as some authors have done, that an electron is both a particle and a wave. It can manifest as both particle and wave, depending on how we observe it. But what it really is, we cannot say. It is neither a particle nor a wave, neti neti. We can see here the relevance of ancient oriental wisdom for modern science. Some would say that in some ways we have gone full circle (see also process language in the following chapter).

Complementarity – The recognition of both/and logic leads to the principle of complementarity. This principle was introduced into quantum physics by Niels Bohr. It means that even contradictory views can complement one another. Thus, the particle view and the wave view complement one another. Both of them together give us a richer view of reality than only one.

On his code of arms, Bohr included the Yin-Yang symbol and the statement "contraria sunt complementa" (opposites are complementary). He and Korzybski extended the complementarity principle to other sciences and life in general. But even today, it is not widely recognized, and it is not much applied in other sciences. Instead, most scientists still use Aristotelian logic and argue whether something is this or that. Rutishauser and I applied the complementarity principle to plant morphology, our area of research (Rutishauser and Sattler 1985, Sattler 2019; see also Appendix 1).

Science and art complement one another, and science and spirituality, "these two domains of human endeavour can and in fact do complement one another" (Dalai

Lama 2011, p. 186). And "apparently opposite terms of One and Many, Form and the Formless, Finite and Infinite, are not so much opposites as complements of each other… not hopelessly incompatible alternatives, but two faces of the one Reality" (Aurobindo, quoted by McFarlane 2002, p. 51).

Fuzzy Logic – Another important new development in modern logic is fuzzy logic. Although the principle of fuzzy logic had been understood for a long time and implied in ancient ways of thinking, such as Yin-Yang thinking, fuzzy logic, in a precise formal way, was developed in the twentieth century. "Fuzzy logic" has at least two meanings. The first meaning is multi-valued logic, which, in contrast to our common two-valued either/or logic, has more than two values. The second meaning is reasoning with fuzzy sets, which was developed by Lofti Zadeh in 1965 (see Kosko 1993). A fuzzy set is "a set whose members belong to it to some degree. In contrast, a standard or nonfuzzy set contains its members all or none. The set of even numbers has no fuzzy members. Each number belongs to it 0% or 100%. The set of big molecules has graded membership. Some molecules are bigger than others and so belong to it to greater degree" (ibid, p. 292).

Kosko (1993) wrote: we live in a fuzzy world. Therefore, "the hard and abrupt contours of our ordinary conceptual system do not apply to reality" (ibid.). **What may appear clear-cut at first sight often becomes fuzzier the closer we look.** Hence one could give endless examples of fuzzy sets: the set of tall men, mean men, aggressive men, violent men, compassionate men, loving women, nasty women, happy people, wise people, law-abiding people, honest people, dishonest people, reliable people, tolerant people, reasonable people, mentally deranged people, friendly dogs, dangerous animals, poisonous plants, interesting discussions, true statements, false statements, indecent remarks, racist remarks, racist actions, terrorists, sanity, insanity, health, science, philosophy, art, spirituality, etc. Kosko (1999) explored fuzziness in politics, science, and the digital age. Although still widely ignored, fuzziness has been recognized, at least in some areas. For example, people refer now to the autism spectrum and other continua.

More examples - If I say, "You are sick" or "You are mentally deranged," this can be harmful and is not encouraging. However, saying "You are a bit sick" or "a bit deranged" is far less negative and may be more appropriate. Imagine how much harm a medical doctor may do who diagnoses his patients in terms of black-or-white thinking! Telling somebody that he is sick and that he has only six months to live is not encouraging and can be harmful. It can even be harmful to tell somebody, "You are healthy" because this is also black-or-white thinking. Who is 100 percent healthy? If a medical test does not reveal a pathology, that does not necessarily mean that the person is "healthy," especially if that person feels a bit sick. In any case, as there are many shades between black and white, there are many shades between health and sickness. It is not only a question of either/or.

Another simplistic black-or-white distinction is the one between moral and immoral people. It can be rather harmful to classify someone as immoral because of one or a few immoral acts. It can also be harmful or at least misleading to classify someone else as moral because who can claim to always be moral? The distinction of moral and immoral people divides. The division creates wounds. Healing thinking such as fuzzy logic avoids such division because it places people on a continuum ranging from moral to immoral.

Although there is some increasing recognition of fuzziness, to a great extent, people and society remain enslaved in the so-called law of the excluded middle. It often happens with regard to the distinction between truth and falsehood. Most people can accept fuzziness between black and white but not between truth and falsehood. The harm that has been done and continues to be done by the refusal to accept truth and falsehood as fuzzy sets seems beyond our imagination. As pointed out by Hoggan (2016), it leads to a toxic state of discourse based on the attitude "I'm Right and You're an Idiot." How often have people claimed that they possess the truth (100 percent), and what atrocities have been committed in the name of that so-called truth? Either/or logic is not conducive to peace because it tends to create the "us versus them" mentality, which may lead to conflict, violence, and war (Chopra 2005). And it does not lead to profound happiness (Chopra 2009).

Healing Thinking – I have referred to healing thinking as the thinking that goes beyond the laws of thought, which involve Yin-Yang thinking, Buddhist, and Jain logic, fuzzy logic, and the principle of complementarity (Sattler 2010). These types of logic can reunite what Aristotelian either/or logic has torn apart, and in this sense, they can be healing. They don't divide people and the world into good and bad and other opposites that provide the basis for antagonism, conflict, and war. They create bridges that heal. Human history has suffered dramatically from the exclusive use of Aristotelian logic. It is time that we use more healing logic to avoid further misery.

> He asked: What is Logic? She said: A mute telling a deaf man that a blind man saw a paraplegic running behind a bald guy while grabbing his hair.

A CONTEMPLATION

Try to describe the face of your beloved or any other person and you will find that it is impossible. You may be able to describe certain features of the face but not the face as it is. It remains indescribable, mysterious, beyond science.

CONCLUSIONS

We use logic – consciously or subconsciously – in everyday life and science. Unfortunately, to a great extent, the majority of people and most scientists still hold on to Aristotelian logic, which can be useful sometimes but remains severely limited. When used where it is not applicable, it leads to distortions, which may have negative consequences such as conflict, aggression, and war.

In contrast, Yin-Yang thinking, Buddhist and Jain logic, complementarity, and fuzzy logic can give us a more realistic picture, can reunite what Aristotelian either/or thinking has torn apart and thus can heal the wounds created by Aristotelian logic, which can lead to greater sanity, better health, more profound happiness and peace. Yin-Yang, Buddhist and Jain logic point even beyond logic and hence beyond science since they also include the indescribable, the mystery of existence, which unites us all.

I find the Yin-Yang symbol especially appealing. In its simplicity, it represents opposites, their complementarity, fuzziness, and, as a whole, the indescribable mystery.

> "Logic means mind. Mind is helpful in understanding the objective world. Mind is a hindrance in understanding the subjective world, because the subjective world is beyond the mind, behind the mind" (Osho).

> "There is something beyond the mind which abides in silence within our mind. It is the supreme mystery beyond thought" (The Maitri Upanishad).

Chapter 6
Language and the Unnamable

As he grew older and wiser, Albert Einstein noted: "In my youth I thought Truth can be known. Now I think otherwise; now I think, Truth is unknowable and will always be unknowable" (Albert Einstein, quoted by Rajneesh 1978, p. 100). Many people, including scientists, might disagree, but Einstein based his assertion on solid evidence: we know about the limitations of our senses due to the organization of our nervous system and sense organs. Some species of animals can sense aspects of reality beyond our direct experience. For example, bees can see ultraviolet patterns in flowers invisible to us; bats can generate and sense sounds beyond our perception. But even if our senses were more truthful, the language, language structure, and logic we use to formulate our experience and insights restrict and distort Truth (that which is). In other words, "**human experience gets filtered and mediated by contingent features of human sensory organs, the human nervous system, and human linguistic constructions**" (General Semantics in Wikipedia). Linguistic constructions comprise words and the language structure and logic that relate words to one another. "Words are probably one of the deepest and most unconscious filters we have" (Falconar 2000, p.VI). "Most of us human beings think that we are masters of words; the truth is they master us, we are enslaved by words" (ibid., p. 3). They master us in a way that tends to remove us more or less from reality because they fragment the wholeness of reality. Through words, "languages have taught us to separate things such as mind and body, time and space, outside and inside (ibid., p. 6). But "there is *no* such thing as an object in absolute isolation" (Korzybski 1958, p. 60/1; see also Korzybski 2010). Everything appears interconnected and integrated into an all-encompassing whole, ultimately the whole universe. Therefore, "to use words to sense reality is like going with a lamp to search for darkness" (Falconar, ibid., p.3). Nonetheless, words can be useful, especially if it understood that "words are not the things we are speaking about" (ibid., p. 60). Words refer to concepts that have been abstracted from the all-encompassing whole. Even the particular objects (fragments) they refer to, they cannot cover entirely. They cover only an aspect because "an object has many characteristics on different levels such as the macroscopic, microscopic and sub-microscopic. Most of these characteristics are unknown to us, and so they are not included in the word we give it, the object's name (ibid., p. 7). Therefore: "**Whatever you might *say* something "is"**, ***it is not***" (Korzybski 1958, p. 409). **There is no identity of the word and the object it refers to. The word (concept) represents only a**

selection of some of the characteristics of the object; it represents an abstraction (abstraction means selection). But "many of us do not see language as a representation of reality, but confuse it with reality itself" (Niebauer 2019, p. 26). Such confusion may lead to various degrees of insanity that usually is not recognized as insanity because it is so widespread.

Let me illustrate this with some examples. The word that refers to the concept "apple" has been defined by some characteristics of apples such as their development, shape, anatomy, etc. It leaves out many characteristics of apples, such as their beauty, brilliance, and interconnectedness with the all-encompassing whole. Thus, Magritte painted an apple and, above the image, he wrote: "Ceci n'est pas une pomme" (This is not an apple).

This may appear puzzling or nonsensical to many people, but a painting of an apple "is" not an apple; it represents only *an image of an apple*, and an image of an apple "is" not the same as an apple; it lacks identity with the apple because an image of an apple cannot include all the characteristics of an apple. Like a word, it cannot cover the whole object. Therefore, "whatever you might say the object "is", *well it is not*" (Korzybski, ibid., p. 35).

Another example: "John is a criminal." Again, John "is" much more than what the definition of a criminal entails. He also has positive emotions more or less similar to those of other human beings.

Words refer not only to single objects but also to categories of objects such as "animal" or "human being." We have to stress that every member of the category, such as every human being, has unique characteristics (Chapter 3). Members of a category are held together only by the characteristic(s) that define it. And often we cannot even find characteristics that apply to all members of the category (see, e.g., Sattler 1986, pp. 82-85). For this reason, one cannot generalize, or generalizations have limitations. Much harm has been done by inappropriate generalizations such as "Men are aggressive."

The Unnamable - Regardless of whether a word refers to an object, or a category of objects, or a category of categories, "reality is far from words and it is very different from what a naïve person thinks it is" (Falconar, ibid., p. 7). Therefore, to come closer to reality, we have to become silent. Instead of using words immediately as we encounter a new situation, it would be helpful to pause and first experience the situation non-verbally. Subsequently, we could use words while recognizing that there is always more than can be said about anything. "Whatever we may say will not be the objective

Science and Beyond

level, which remains fundamentally un-speakable (Korzybski, ibid., p. 34). Thus, **the recognition of the non-identity of word and object, language and reality, leads to the recognition of that which is beyond language, the unnamable, the mysterious, the unknowable, which may be experienced in silence and meditation.** Such recognition can lead to greater sanity, better health, more profound happiness and peace (see Appendix 2).

Since science uses language and mathematics, a form of language, it cannot reach the unnamable, the mystery beyond language (Sattler 2016). **However, science can point to the unnamable without ever reaching it.** Korzybski's Structural Differential is a device that points to the unnamable – he referred to it as the un-speakable, ultimate reality beyond words.

Korzybski's Structural Differential - Although the unnamable transcends language, language remains, of course, important for communication. And although words and language appear far from reality, they have some connection with reality because they have been abstracted from reality, which means that they contain some selected features of reality. Korzybski devised the Structural Differential to indicate in a symbolic form **the process of abstraction, which means selecting features from reality.** In this process, different levels of abstraction can be distinguished: objects or events, sensation, description, and inferences.

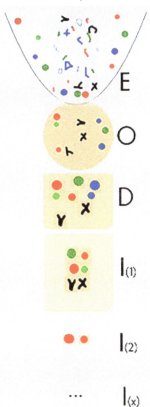

The diagram shows Steve Stockdale's (2020) representation of Korzybski's Structural Differential. E represents an object or an event. Our sensory experience (O) of that object or event includes an impoverished aspect due to the limitation of our sense organs and nervous system. Then, the description of our sensory experience (D) is even more limited (even more limited than indicated in the diagram). And when we draw inferences ($I_{(1)}$, $I_{(2)}$, $I_{(x)}$) from our description, further limitations ensue. Thus, we lose more and more of the object or event as we proceed from sensation to description to inferences. In the above diagram, this loss is indicated in a simplified symbolic way. An object or event does not consist of separate traits. Traits are also formed through abstraction.

Let's take a flower as an example. It represents an object or an event. Our sensory experience of it abstracts (selects) certain features from it due to the limitations of our sense organs and nervous system. So, we lose part of the reality of the flower. For example, we lose ultraviolet patterns because we cannot sense ultraviolet. Then, when we describe the flower using language, we abstract (select) again because language cannot fully represent our sensory experience. For example, we may say that the flower is red, but we miss the finest nuances in colour and texture. Finally, when we draw inferences about the flower, we abstract further when, for example, we state that the flower represents a modified shoot. If that is all we see in a flower, it is far removed from the real object or event of the flower, although it still contains some information about the flower.

Another example that demonstrates how much we lose through the process of abstraction is a sunset. Even a poet could not describe the reality of a sunset because even a poetic description would be only an abstraction, an aspect of that reality.

Photo of a sunset. Courtesy of Rory Skelly.

Korzybski's famous saying: "**a map *is not* the territory it represents, but, if correct, it has a *similar structure* to the territory, which accounts for its usefulness**" (Korzybski, ibid, p. 58). We can have different maps that refer to the same territory. For example, we can have morphological, geological, economic, ethnological, and many other maps that refer to a country like Canada. None of these maps "is" Canada, all of them function as abstractions from Canada based on different selections of characteristics. They complement one another. Hence, the importance of complementarity (see Chapter 5). In science and philosophy, we often forget that theories and philosophies function like maps that may be complemented by different, even contradictory maps as long as these maps are based on evidence (see Chapter 7).

"During English class one morning, Miss Goodbody calls out,
"Betty, tell me the meaning of the word 'trickle.'"
"To run slowly," says Betty.
"Quite right," says Miss Goodbody. "Now tell me the meaning of the word 'anecdote.'"
"A short funny tale," says Betty.
"Good girl," says Miss Goodbody. "Now, Lucy, see if you can give me a sentence with both those words in it."
Lucy thinks for a moment: "Yes, I know," she says. "Our dog trickled down the street wagging his anecdote." "(Osho 1998, p. 132).

Harmful Language and Healthy Language - Harmful language and thinking confuse abstractions with the objects or reality from which they have been abstracted. In contrast, healthy language and thinking are based on an awareness of abstraction, an awareness of the non-identity of map and territory, word and object, language and reality. Hence, healthy language and thinking go beyond the law of identity in Aristotelian logic. They do not succumb to **the myth of identity** that has deeply infected our society. Most, if not all, conflicts and wars appear to be based on or related to this myth, the lack of awareness of abstraction. If we think that the other person or nation "is" evil, and if we want to eradicate evil, we feel that we have to fight or kill. If, however, we recognize that that person or nation is evil and good and infinitely more than we can express in words, then we can become silent and also connect to the goodness. Jampolsky (1979, pp. 124-125) recounts how one night, he was called to see a patient on the locked psychiatric ward. Through the small window in the door of the patient's room, he could see that the patient had become violent and aggressive. He felt scared and afraid to enter the patient's room. However, as he continued to look through the window, it occurred to him that despite his forceful and aggressive behaviour, the patient also seemed scared. Then, admitting to each other that they both felt scared created a bond, and as a result, Jampolsky could walk into the patient's room, talk to him and give him medicine without getting hurt. If he had seen in him only the obvious aggressiveness, if he would have simply labelled him an aggressive patient, he could not have treated him peacefully. Kierkegaard wrote: **"Once you label me, you negate me."**

One may be theoretically aware of abstraction and yet forget it in practical situations. To help us remember, Korzybski devised extensional semantic devices. One of these devices involves adding "etc." when we use "is" (or "are"). For example, instead of saying, "Jampolsky's patient is aggressive," one would say, "he is aggressive," etc., which includes his other traits and indicates that his aggressiveness is an abstraction. Furthermore, instead of referring simply to Jampolsky's patient, one would refer to Jampolsky's patient-October 22, 2009-universe, which indicates the date of the encounter and the context that extends into the universe. Including the date indicates when the patient was aggressive; at another time, he may not have been aggressive. Hyphens are used to emphasize interconnectedness such as, for example, the patient's interconnectedness with his environment, which may include the whole universe. Quotation marks are used to indicate the highly abstract nature of a word such as, for example, "love," which can have many different meanings. Korzybski pointed out how the use of the extensional devices can be healing and thus lead to greater sanity (for a more detailed discussion of the extensional devices, see the section on Speech in Appendix 2: Health and Sanity of Body, Speech, and Mind).

You might find these extensional devices contrived or awkward. You don't need them if you always remain aware of abstraction. But who, except perhaps some rare individuals, can claim to always retain an awareness of abstraction? Therefore, we need training in the use of extensional devices. With my best intentions, I doubt that I always applied them in this book. Common habits of writing appear very strong. Especially the use of the "is" of identity seems deeply ingrained in most of us, including myself, which means that we tend to forget the "etc."

Instead of using the extensional device "etc." that still includes "is" such as "He is sick, etc.," one could use constructions that do not include "is" (see Kodish and Kodish 2011, Kellogg 1987, Kellogg and Bourland 1990/91). For example, instead of saying, "He is a politician," which includes what has been called the "is of identity," one could say, "He works as a politician." He "is" much more than just a politician. And instead of saying, "The rose is red," which includes what has been called the "is of predication," one could say, "The rose looks red (to me)." It "is" infinitely more than just red. Since the "is of predication" also involves an identification, it could be subsumed under the "is of identity" in the widest sense. The "is of identity" represents the law of identity, one of the three laws of thought that characterize Aristotelian logic (see the preceding chapter). Korzybski was keenly aware of the inadequacy of the laws of thought in Aristotelian logic and therefore referred to his philosophy as "Non-Aristotelian Systems and General Semantics" (Korzybski 1933/58). Non-Aristotelian Systems include Aristotelian logic but go far beyond it.

We have been deeply conditioned by a language structure that implies the law of identity, and therefore identity and identification remain a major issue for many people, groups, nations, etc. And "every identification is bound to be in some degree a misevaluation (Korzybski, ibid., p. XXXIV) that may lead to conflict and possibly even war (see also Kodish and Holsten 1998). Complete identity exists only in mathematics when we say A=A. In the real world, we have at best approximations to identity, but not complete identity because everything changes, as Heraclitus pointed out before Aristotle published his Logic. Although my name is Rolf Sattler for my whole life, the Rolf Sattler of yesterday is not identical with the Rolf Sattler of today.

It happens again and again that "we read unconsciously into the world the structure of the language we use" (Korzybski, ibid., p. 60). But even if we become aware of the importance of language structure, we may not fully realize to what extent it affects our psyche. Korzybski emphasized that "psychological" can mean "psycho-logical," which underlines the importance of logic and language structure for our psyche. He also referred to semantic reactions. If we use the "is" of identity and predication, semantic reactions may lead to various degrees of insanity, a delusion about reality (see also Appendix 2). For example, if we claim that this person or this nation is evil, such a statement may lead to paranoia, insanity, violence, and even war. Remember former American president George W. Bush referring to the "axis of evil" and the subsequent invasion of Iraq. "A great deal of our human problems, confusions, conflicts, and violence, at diverse levels - personal, interpersonal, societal, and international - can be attributed to our use of "is" in the identification and predication mode" (Dawes 2010).

Identification and a lack of awareness of abstraction can also lead to "a tendency to make static, definite, and, in a way, absolutistic one-valued statements. But when we fight absolutism, we quite often establish, instead, some other dogma equally silly and harmful. For instance, an active atheist is psycho-logically as unsound as a rabid theist" (Korzybski, ibid., p. 140). When we forget the characteristics left out in the abstraction, we tend to think that we are right and that our statement is the only possible one.

Process Language – Besides implementing Korzybski's extensional devices, avoiding noun-verb structure could also reduce distortions and harm and bring language closer to reality. Reality appears profoundly dynamic, but this dynamic is obscured by the noun-verb structure of most languages. Nouns (and pronouns) involve and refer to separate entities, and verbs to processes. Separateness implies static because although an entity (such as a person) may change, it remains a separate entity. Since nouns precede verbs, they appear primary, processes secondary. Thus, this kind of language structure reflects and reinforces a worldview according to which the world consists of separate things, persons, ideas: you and I appear separate, the trees appear separate from one another and us, the good appears separate from evil, etc. Hence, this language structure leads to alienation. However, holistic science and personal experience, such as the mystical experience of oneness, reveal the world as basically whole and one. David Bohm (1981) referred to **undivided wholeness** and to **holomovement** to underline the dynamics. **In this view, process appears primary, and entities are seen as abstractions from the process.** The question then arises: can we develop a language that reflects the fluidity of manifest reality? In other words, can we develop a process language? Since process is represented by the verb, we would want a language in which the verb, not the noun or pronoun, plays the primary role. In *Tlön, Uqbar, Orbis Tertium,* Jorge Luis Borges gave a nice example of process language. Instead of saying: "the moon rose above the river," he suggested: "upward behind the onstreaming it mooned." In *Fragmentation and Wholeness* (1976), Bohm attempted to construct rudiments of a process language, which he called the rheomode. I am not aware of further developments in this respect, and so far, I have not been able to construct a purely verb-based language that satisfies the linguists I consulted. However, I have learned that at least some Amerindian languages appear much more verb-based than English. In these languages, the verb appears central and primary. Spielmann (1998, p. 46) wrote: "The verb is the heart of the Ojibwe language." According to Whorf (2012, p. 310), there are even native **languages such as Nootka that consist only of verbs.** "Sakej Henderson said that when talking in, say, Mikmaq, he could talk all day long and never utter a single noun" (Dan "Moonhawk" Alford et al. 2009). I have also learned that instead of saying "I love you" (subject-verb-object), a Japanese would simply say "Aishiteru," which means "loving": the activity, the process of loving. On this view, lover and beloved, subject and object, are seen as abstractions from this process (Berendt 1991, p. 46). I think that if we begin with such changes in language structure and then extend them to more and more situations, we might be able to overcome much alienation, which could lead to a more harmonious and peaceful world.

I read that during the early evolution of human language, nouns and verbs were not always clearly differentiated from one another (Deutscher 2005, Hurford 2014). One can see then "that objects are also events, that our world is a collection of processes rather than entities" (Watts 1957, p. 5). Also, "Chinese makes no rigid distinction between parts of speech. Nouns and verbs are often interchangeable, and may also do duty as adjectives and adverbs" (Watts 1975, pp. 8-9). If the noun functions as a verb, a sentence with a noun becomes purely verb-based and thus reflects the fluidity of reality. For example, if the noun "flower" functions as a verb indicating process, it means "flowering," and then the sentence "The flower appears red" means "Flowering appears red," which means that the process of flowering appears red. One could also just refer to "Red flowering." I was told that, in ancient Hebrew, the word "flower" actually meant "opening," thus also indicating process.

Dan "Moonhawk" Alford et al. (2009) and others pointed out that **the primacy of process in some native languages resonates well with quantum physics**, which also emphasizes the primacy of process. They mentioned that Heisenberg "lamented the limitations of noun-heavy Western languages in explaining physics." Thus, we have come full circle from native languages and their associated dynamic worldview to the dynamic worldview of modern quantum physics. Both appear more realistic than the mechanistic materialist worldview of Newtonian physics and our mainstream culture and science. Both appear well suited to counteract the alienation that has been created by the mechanistic materialistic worldview. In a world where we do not appear divided and separate from each other, we can feel oneness and compassion.

Besides quantum physics, the primacy of process has also been emphasized in other sciences (Nicholson and Dupré 2018) and in process philosophy, as elaborated by Whitehead (1929). In plant morphology, my field of specialization, I developed a process morphology according to which plant structures are seen as processes that arise out of the unnamable mystery (see Appendix 1).

A MEDITATION

With eyes closed, suddenly and abruptly stop breathing and moving. You may notice that all thoughts and words are gone immediately. Thus, the unnamable, mysterious, beyond science.
(for more on STOP meditations, see Osho 2010).

CONCLUSIONS

As Korzybski has shown through his Structural Differential, language abstracts from reality and thus filters out much of the richness of reality. Assuming that language represents the world as it is implies a profound misconception, a delusion – hence, a degree of insanity, as Korzybski has emphasized so much. Language can represent at best aspects of reality, but not reality as it is. If Truth is understood as that which is, language cannot represent Truth, at best aspects of Truth. Truth remains unnamable, beyond language and science that uses language.

Healthy and healing language recognizes the process of abstraction, whereas harmful language implies a mistaken belief in the identity of map and territory, word and object, language and reality. Extensional devices, suggested by Korzybski, aid in a healthier use of language. Process language that emphasizes the verb can counteract fragmentation that leads to alienation.

Language, its structure and logic affect our psyche: they are psycho-logical. If children and adults were taught the process of abstraction and the use of extensional devices, we would live in a saner world.

> "Everything I say is limited, limited, limited" (Alfred Korzybski).

> "We are prisoners of our language" (Alberto Melucci).

> "The ultimate Reality is Nameless" (Franklin Merrell-Wolff).

> "Existence is beyond the power of words to define: terms may be used but are none of them absolute" (*Tao Te Ching*, Bynner's translation).

> "Every great teacher has said that what is really True and Real cannot be expressed in words or grasped by the rational mind" (Ravi Ravindra).

Chapter 7
Empiricism and Just Sensing

According to empiricism, science has an empirical basis that resides in shared sensory perceptions, called facts. Hence, empiricism is based on facts. Sensory perception can be understood as a description of sensations. As pointed out in the preceding chapter, sensations are abstracted from reality. Due to the limitations of our sense organs, we sense only a restricted aspect of reality, and as we describe our sensations using language and concepts, we represent only aspects of our sensations. Thus, as pointed out in the preceding chapter, **sensations and their description, that is, sensory perception, remove us to some extent from reality.** Failing to recognize this constitutes a degree of insanity (see Appendix 2). **If Truth is understood as that which is, sensory perception does not represent the Truth, only an aspect of the Truth. Religious or spiritual experience and the arts can reveal other aspects of the Truth, which may reach even deeper than sensory perception.**

This painting by the surrealist painter René Magritte transcends sensory perception.

The limitations of empiricism have been known for a long time. The German philosopher Immanuel Kant emphasized that we cannot know what he called "things-in-themselves," which means that we cannot know reality as it is. We perceive reality in terms of categories such as space and time. Thus, the mind (reason) shapes and structures our experience. This recognition bridges the impasse between rationalists who believe that we can know reality through reason alone and empiricists who think that we know reality through sensory experience.

In meditation we can go beyond the mind, the thinking mind, and its categories and concepts. You may ask: Why go beyond the mind? Because we have not only a mind, we also have a heart and gut. And upon closer inspection, we can find out that the heart and the gut have a greater depth than the mind. Therefore, spiritual traditions often emphasize more the heart and the gut than the mind. But the integration of the gut, the heart, and the mind seems most desirable.

In empiricism, sensory perception is normally limited to shared sensory perception of the outer world. In contrast, **radical empiricism** that was championed by William James (1912/1976) **is based on "pure experience,"** which includes "the immediate flux of life which furnishes the material to our later reflection with its conceptual categories" (James 1912/1976). Thus, pure experience does not seem to be restricted to outer experience but **also includes inner experience and may comprise all experience - sensory, mental, and spiritual experience – and thus enlarges the scope of empiricism enormously.**

Wilber (2001) made the distinction between **narrow and broad science**. Narrow science is based on sensory perception of the outer world, whereas broad science is based on all-inclusive pure experience. However, in as much as pure experience is conveyed through language, it cannot reach reality (ultimate reality) because language abstracts (selects) from reality and therefore can at best represent only an aspect of reality, as I pointed out in the preceding chapter. And thus, broad science is also limited because it implies language.

Goethe developed a "**delicate empiricism**" ("zarte Empirie") that relies on empathy, intuition, and imagination to promote a participatory engagement with nature (see, for example, Kirchoff 2002, Wahl 2017). Like James's radical empiricism, it also reaches deeper than empiricism as it is usually understood. But it has limitations because it uses language and concepts to describe the deeper experience.

Just Sensing - Contrary to sensory experience that implies language and thought (concepts), **sensing implies a direct experience that does not involve language and thought** (Sattler 1986, p. 70). Although limited due to the limitations of our sense organs, paradoxically, it may open vistas into the infinite. People who are able to completely immerse themselves in a sensual experience without any interference of language and thought would agree. William Blake said: "If the doors of perception were cleansed every thing would appear to man as it is, infinite." Cleansing the doors of perception means removing the overlay of thought and language on our sensations, allowing ourselves the deepest immersion into the natural world without the interference of preconceived ideas.

One may ask: Since our sense organs are limited, how is such a vision of the infinite possible? Maybe the following analogy will be an answer. We can look through a narrow window at the sky, the narrow window representing our limited sense organs and the sky the infinite. Thus, even through a narrow window, we may have access to the infinite. Just sensing is not about gaining knowledge; it involves the awareness of being one with the infinite.

Even in our cerebral culture, which stresses the thinking mind very much, sensing without the interference of the mind and language may be possible. Artists may be more receptive to it. It is also emphasized in various body-oriented healing arts and learning, such as the Feldenkrais Method and Madelyn Kent's Sense Writing, which involve awareness of body sensations (Kent 2020).

In stillness all our senses can be a door to the infinite. A spontaneous opening may bring us there. Relaxation and various methods of meditation can be helpful. Kelly (2015), in his book *Shift into Freedom*, suggested several ways how our senses can give us glimpses of the infinite. For example, instead of looking out into the environment, let it come into us. Practice seeing as receiving (ibid., p. 115). Or let a sound or scent or taste or touch transport you into the infinite.

Since sensing does not involve thought and language, it goes beyond science, which requires the mind and language. Besides sensing other ways of direct experience can reveal profound in-sights into the infinite such as **"theoria" in Plato's non-dual vision** (that is much less known than his dualistic view). "Theoria: 'theon oro' – which means, I see (oro) the Divine (theon). Theoria means ascending to the View of the Divine, where the veil of ignorance drops and the cosmic play is revealed as what it is: the miraculous, mysterious, ever indescribable play of Oneness…In and from that view, what I *see*, as *experiencing*, or *experience as a form of seeing*, is the stripping away of all knowledge, all science" (Potari 2018, p. 13). It leads "beyond subject and object duality, to a view of Being as consciousness-aware-of-itself" (ibid.). Hence this view of theoria, which is beyond science, differs very much from the modern scientific understanding of a theory testable by facts.

Testability of scientific hypotheses, theories, and laws constitutes a strength of science. Through tests, scientific tenets can be confirmed or disconfirmed. What is not testable is usually excluded from science. However, what may be untestable at a given time may become testable later on. Therefore, excluding it from science may hinder scientific progress.

Hypotheses, theories, and laws are tested through facts. Normally, facts are observed in the external world through our senses, especially vision and hearing. Pure experience, as noted above, enlarges the scope of facts enormously. For example, it also includes extra-sensory perception. In any case, facts are of utmost importance. If the facts do not support a theory, according to empiricism, the theory should be discarded or modified so that it can accommodate the contradictory facts. However, as has been pointed out by Thomas Kuhn (1962) and others, science does not always work this way. Progress in science is not simple and straightforward. In Chapter 2, I pointed out that facts are theory-laden and value-laden, and in Chapter 4, I indicated how facts can be selected and manipulated through the expectancy effect, the observer and experimenter effect, selection bias, confirmation bias, consensus bias, reiteration bias, and publication bias. Furthermore, contradictory facts are often ignored, declared erroneous, or explained away by *ad hoc* hypotheses so that scientists can continue to hold on to their preferred theories and laws.

> The Trump Travel ban was refused due to lack of evidence. Apparently, "I know it, you know it, everybody knows it" wasn't enough.

***Ad hoc* Hypotheses** - Usually *ad hoc* hypotheses are introduced to save theories, paradigms, or world views from contradictory evidence, in other words, to explain away the contradiction. For example, telepathy appears to contradict the materialistic worldview. To retain the materialist worldview, an *ad hoc* hypothesis that is often used claims that the evidence for telepathy is based on faulty methodology (which is not generally the case). It seems that almost any theory, paradigm, or worldview can be defended through *ad hoc* hypotheses. However, as more and more contradictions accumulate, eventually the status quo may be given up. But this may take a long time and may happen only after the death of its defenders. As Max Planck, one of the founders of quantum physics, noted: "A scientific truth does not triumph by convincing its opponents and making them see the light, but rather because its opponents eventually die and a new generation grows up that is familiar with it." Schopenhauer noted three phases in the acceptance of a revolutionary new idea: 1. It is mocked, 2. It is dismissed, 3. Its opponents say: It is obvious; we knew it all along. They cannot honestly admit how much they were opposed. So much for the honesty and rationality of science.

Ad hoc hypotheses are not necessarily incorrect, but their prime purpose is to protect a theory, paradigm, or worldview from contradictory evidence so that their defenders, who have much invested in that particular theory, paradigm, or worldview, can continue to hang on to them. For example, one reason why the materialistic and mechanistic worldview still persists today, especially in the life sciences, is that it has been surrounded by a belt of *ad hoc* hypotheses. **Because of the use of *ad hoc* hypotheses, it has been** said that "any new evidence can, with sufficient effort, be made to fit a preexisting paradigm [or theory or worldview]" (Eisenstein 2013, p. 3). How much, then, does science differ from an ideology or religion? (see Chapter 9: A Radical View of Science).

Confirmation Holism and Ontological Relativity - Quine (1980) pointed out a network of interconnected tenets. If any particular empirical evidence clashes with one of these tenets, changes can be made within the network to accommodate the evidence. Thus, confirmation occurs within the whole network (not just with regard to one particular statement or tenet within the network). As long as we are willing to make changes somewhere within the network, contradictions can be resolved. Thus, no particular empirical evidence can force the revision of a tenet such as a theory if changes are made elsewhere within the network. On the other hand, any part within the network, such as a theory, can be changed as a result of new empirical evidence. According to Quine, theories are under-determined by empirical evidence since there could be more than just one theory to account for any empirical evidence. Consequently, we have to acknowledge considerable ontological relativity, which means that any collection of empirical evidence can be accounted for by more than just one theory. However, whether this is always the case has been debated. According to partial confirmation holism, it is not always so.

> It has been said that Egyptians invented cement . . . Archeologists examined ruins for evidence but there is nothing concrete.

Simplification seems unavoidable in scientific methodology and it leads to distortions of reality. As Bart Kosko put it: "Scientific claims or statements are inexact and provisional. They depend on dozens of *simplifying* assumptions and on a particular choice of words and symbols and on "all other things being equal."... **When you speak, you simplify. And when you simplify, you lie**" (Kosko 1993, p. 86; see also *Science's First Mistake* by Angell & Demetis 2011).

> "The village priest approaches a group of small boys sitting in a circle around a dog.
> When he comes up to them, he asks, "What are you doing to the dog?"
> Little Ernie answers, "Whoever tells the biggest lie, wins the dog."
> "Oh, dear," exclaims the priest, ", I'm surprised at you boys.
> When I was young like you, I never told a lie."
> There is silence for a while, until little Ernie shouts out, "Okay, give him the dog!" " (Osho 1998, p.107).

Power-Knowledge – It seems naïve to believe that science is only based on facts. Individual scientists and scientific communities, such as the conservative medical establishment, use their influence and power to uphold their views despite contradictory evidence. Thus, **power shapes science to a considerable extend.** In this regard, Foucault coined the expression "power-knowledge," which means that, **at least to some extent, power determines or influences how we gain and maintain knowledge** (Foucault 1983). **What is presented as objective truth appears contingent on historically conditioned forces that involve power.** For example, materialist scientists and the community of materialist scientists of mainstream science control to a great extent research in science. We often hear that science is evidence-based, which gives the impression that evidence determines objectively which theories are supported. However, powerful materialist scientists and their community control which evidence is acceptable. Evidence that appears incompatible with a materialist worldview is ignored, dismissed or suppressed as misinformation, disinformation, non-science, pseudoscience, or conspiracy theory (see below). For example, evidence for telepathy and other psychic phenomena is rejected. There are, of course, scientists who can transcend the materialist strait jacket. But mainstream science remains much under the control of materialist scientists and their community. In medicine, the powerful medical establishment that espouses materialism controls which evidence is acceptable. Thus, much alternative medicine is rejected because it does not conform to the materialist worldview. Again, there are pockets of resistance that go beyond materialism, but they are kept under control. The powerful medical establishment conspires with the pharmaceutical industry and government to suppress alternative and integrative medicine as much as possible (Lanctôt 1995, Edwards 2007, All 2012). Unfortunately, this was also the case during the COVID-19 pandemic, and as a result, many people died whose lives could have been saved (see Chapter 9). We could see how, **besides evidence, politics, social and psychological factors play an important role.**

Peer Review – Although peer review can improve the quality of publications and prevent the publication of poorly executed research, it can also enhance and protect power-knowledge. This is how it works: A scientist submits a paper for publication to the editor of a journal or publishing house. Then the editor sends the paper to reviewers – peers who are scientists in the same field of specialization – to have the paper evaluated. The reviewers recommend the paper for publications, maybe with recommendations for improvement, or reject it. If the editor does not like the paper and therefore does not want to publish it, he or she may select reviewers who will reject the paper. Most editors know the reviewers well enough to predict with some confidence who will accept or reject the paper. In this way peer review can be used to consolidate the editor's bias and power. And inasmuch as the editor supports the view or dogma of a group or community of scientists, (s)he can maintain and reinforce their power-knowledge. Unfortunately, the general public seems largely unaware of the potential manipulation through peer review and is often misled into believing that a peer-reviewed paper is always of superior quality or even "true" (see Peer Review in Wikipedia).

Pseudoscience - In contrast to science based on empirical evidence, pseudoscience lacks an empirical basis. One should think then that it is easy to distinguish pseudoscience from science. But it is not so because, as I pointed out already, it is not always clear what counts as evidence. Furthermore, scientists and the scientific community often use their power to discredit contradictory evidence and tenets. One way of doing this is to declare such evidence and tenets as pseudoscience. Consequently: "The term 'pseudoscience' has become little more than an inflammatory buzzword for quickly dismissing one's opponents in media sound-bites" (McNally 2003). Such dismissal can have harmful consequences. For example, alternative medical practices such as homeopathy or acupuncture that are not based on materialist mainstream science are often dismissed as pseudoscience. In this way, the uneducated public is misled and may be robbed of beneficial treatment. Even government officials are often misled by the power-knowledge of the materialist scientific community that still dominates mainstream science (see Chapter 9).

> "Paddy is very, very ill indeed, so Maureen sends for Doctor Gasbag.
> After a brief examination, the doctor announces that Paddy is dead.
> "I am not," says Paddy from his bed.
> "Be quiet," says Maureen. "Do you think you know better than the doctor?" "(Osho 1998, p. 511).

Conspiracy Theory – Nowadays, the term conspiracy theory is increasingly "deployed against anyone who questions authority, dissents from dominant paradigms, or thinks that hidden interests influence our leading institutions. As such, it is a way to quash dissent and bully those trying to stand up to abuses of power" (Eisenstein 2020b). "Labelling someone a conspiracy theorist, while studiously avoiding any discussion of the evidence they highlight, is extremely common in the mainstream media" (Davis 2020). Thus, the label "conspiracy theory" has become one of the most powerful weapons to exercise the power-knowledge of the scientific establishment. "This weaponised and derogatory term was invented by the CIA in the 1960s to discredit and defame those who questioned the truth of the official Warren Commission report on the Kennedy assassination… many journalists are cowed by the threat of being branded in this way, even if this is entirely unjustified. Judgement is a matter of reason, evidence, and discrimination as each of us join the dots and proceed to draw our own conclusions. The key issue is which dots to join and on what basis" (Lorimer 2020; see also Davis 2020). During the COVID-19 pandemic, much helpful evidence was dismissed and suppressed by labelling it conspiracy theory (see Chapter 9).

Anecdotal Evidence - Another way to dismiss evidence is to declare it anecdotal. But just because something happened only once or occasionally does not mean that therefore it has not occurred. In science, even rare and exceptional events should be considered. If they are dismissed, we lose the opportunity for further investigation that might lead to a revision of the status quo. But mainstream science seems to be more interested in retaining its power-knowledge.

The dismissal of facts as anecdotal appears widespread in mainstream medicine because the conservative medical establishment does not want to have its power-knowledge challenged. But there are many reports of people being healed through methods of alternative medicine. If these cases would be further investigated, much could be learned for the benefit of humanity. Many of these cases are well documented. For example, Norman Cousins suffered from ankylosing spondylitis, a disease of the connective tissue that was considered incurable by mainstream medicine. In his book *Anatomy of an Illness* (1981), he documented how he cured himself through laughter and Vitamin C. Nonetheless, skeptics often remain skeptical. Although skepticism seems desirable, a problem with many skeptics is that they are skeptical towards others but not towards themselves.

The Semantic View of Scientific Theories and Laws - An important innovation in the understanding of science is the semantic view of scientific theories and laws. According to this view, as understood by Giere (1979), theories and laws are considered definitions. Thus, we no longer ask whether they are true or false. Instead, we ask whether they apply to particular situations. If more than one theory or law applies, they are considered complementary (see also Chapter 5).

For example, Newtonian physics or Mendelian genetics apply to many situations and therefore constitute useful theories and laws. We know, however, that they don't apply to all situations, and consequently they are not generally true. But according to the semantic view, the question is no longer whether they are true. The question is whether they can be applied. And where they can be applied, they have explanatory and/or predictive power.

The semantic view has many advantages (Giere 1979). It recognizes that truth remains elusive in scientific theories and laws. As pointed out in Chapter 2, proof appears unattainable in science, whose strength resides in its open-endedness, which means: no final word, no assurance of the ultimate truth. We have to become humbler, which can have many beneficial consequences.

A MEDITATION

Close your eyes and smell your favourite perfume or use any other of your senses to enter infinity beyond language, thought, and science.

CONCLUSIONS

According to empiricism, science has an empirical basis that resides in shared sensory perception. Contrary to naïve realism, sensory perception does not represent reality as it is. It is abstracted (selected) from reality, which means that it does not represent the Truth if Truth is understood as that which *is*. At best, sensory perception represents aspects of the Truth. Believing that sensory perception and empiricism give us the Truth leads to a degree of insanity because this belief is not founded in reality. To me, insanity means being deluded about reality.

Radical empiricism is based on pure experience, which is not restricted to outer experience but also includes inner experience. Hence, pure experience comprises all experience - sensory, mental, and spiritual experience – and thus enlarges the scope of experience and science enormously. However, in as much as pure experience is conveyed through language, it cannot reach reality (the unnamable) because language abstracts (selects) from reality and therefore can at best represent aspects of reality, as I pointed out in the preceding chapter.

Broad science, although based on pure experience, is also limited because it too, implies language. Goethe's "delicate empiricism," which relies on empathy, intuition, and imagination, also relies on language. However, beyond language, the arts, religious or spiritual experience and just sensing without the interference of the mind and language, may open the door to the infinite beyond science and thus lead to the awareness that "you are the universe" (Chopra and Kafatos 2017). "This profound knowledge isn't new. In ancient India, the Vedic sages declared *Aham Brahmasmi*, which can be translated as "I am the universe' or "I am everything." They arrived at this knowledge by diving deep into their own awareness" (ibid.). Moments that appear too difficult to bear may be transformed through this awareness, which may arise spontaneously, in deep meditation, contemplation, and through other means such as, for example, sacred acoustics (Newell et al. 2020)

"Theoria," as understood in Plato's nondual vision, also transcends science as it reveals "the miraculous, mysterious, ever indescribable play of Oneness…In and from that view, what I *see*… is the stripping away of all knowledge, all science" (Potari 2018).

In contrast, theories, as understood in science, are supported and tested by facts based on sensory perception. However, if a fact or facts contradict a theory, that does not mean necessarily that the theory will be modified or abandoned. There are various ways how scientists resolve the contradiction without sacrificing the scientific theory. They may simply ignore the contradictory fact(s) or declare them faulty, or they may invent an *ad hoc* hypothesis that saves the theory, or they may invoke confirmation holism, according to which any theory may be retained as long as adjustments are made elsewhere in the network of tenets.

According to empiricism, science is based on empirical evidence. However, besides empirical evidence, power such as the power of powerful individual scientist, organizations, communities, etc. plays an important, often decisive role that may even override empirical evidence. This leads to "power-knowledge" that shapes science to a considerable extend. It restricts deviations from the status quo. What does not fit is often declared misinformation, pseudoscience, or conspiracy theory.

According to the semantic view of theories and laws, the question is no longer whether theories and laws are true, but only whether they apply.

> Simplification seems unavoidable in science and leads to distortions of reality. "And when you simplify, you lie" (Kosko 1993).

"Reality transcends theory, reason, and observation" (McFarlane 2002).

"To sense that behind anything that can be experienced there is a something that our mind cannot grasp and whose beauty and sublimity reaches us only indirectly and as a feeble reflection, this is religiousness. In this sense I am religious" (Albert Einstein).

> "If the doors of perception were cleansed every thing would appear to man as it is, Infinite. For man has closed himself up, till he sees all things thro' narrow chinks of his cavern" (William Blake).

Chapter 8
Mechanistic Materialist Science, Holistic Science, and Beyond

The limitations of science pointed out so far appear unavoidable because they are at the basis of any scientific approach. Thus, there are limitations to explanation and prediction (Chapter 1). Since we cannot foresee the results of future observations and experiments, proof remains unattainable (Chapter 2). Replication, which plays a fundamental role in scientific methodology, has limitations because of the uniqueness of events and changing context (Chapter 3). Objectivity has limitations for a variety of reasons (Chapter 4). Since science uses language and logic, it remains constrained by the unavoidable limitations of language and logic, although less used kinds of language and logic could make a great difference (Chapters 5 and 6).

The empirical approach reflects the limitations of our experience (Chapter 7).

Mechanistic materialist science - In contrast to these unavoidable limitations, mainstream science, as it embraces materialism and mechanism, imposes unnecessary limitations because science need not be materialist and mechanistic. Materialism and mechanism represent a worldview that we inherited from the Age of Reason (also called "Enlightenment," although it does not appear enlightened). According to materialism and mechanism, reality is a material mechanism. To a great extent, mechanistic materialism has become a deeply rooted dogma in mainstream science and mainstream culture (Sheldrake 2012/20, Beauregard et al. 2014). According to this dogma that seems especially pernicious in biology and medicine, it is taken for granted that an organism is a physical mechanism, like a complicated machine that consists of interacting material components. As in a malfunctioning machine, sickness then is understood as the malfunctioning of one or more of the components of a sick person, and healing requires repairing or replacing the deficient component(s) such as an organ like the heart or kidney. Of course, some medical doctors can see beyond mechanism, but the majority of the medical establishment seems to be still caught in mechanistic materialism.

> "Sally Goldberg goes to the Doctor to ask for some help in losing weight before her wedding day.
> He prescribes a course of slimming pills for her.
> A few days later she returns to his office.
> "These pills have awful side effects," she says, worriedly. "They make me feel terribly passionate and I get carried away. Last night I actually bit off my boyfriend's ear."
> "Don't worry," says the doctor., "an ear is only about sixty calories." (Osho 1998, p.344).

Mechanistic materialist science appears deeply embedded in our materialist mainstream culture of capitalism, consumerism, and militarism, and for this reason, it seems very difficult to go beyond mechanistic thinking. Mechanistic materialist science, capitalism, consumerism, and militarism reinforce one another. They have led to the exploitation and ruin of our environment and thus have greatly contributed to the ecological crisis and the disenchantment of the world. To create a saner, happier and more peaceful world we have to go beyond mechanism and its reinforcement through materialism, consumerism, and militarism (see Gorbachev 2020, Tudge 2021. and Appendix 2).

One way to go beyond materialism is to recognize the spiritual in matter; in other words, not to divide reality into matter and spirit. But what normally is considered matter as in materialist science, has been divorced from spirit. This is contrary to many indigenous traditions in which nature is not fragmented into matter and spirit, and therefore nature is seen as spiritual and sacred. To avoid reference to spirit, which may create misunderstandings because it can have different meanings, I prefer to refer to the unnamable mystery (Chapter 6).

Further restrictions of materialism - Within materialist science, one can find further restrictions that lead to a further narrowing of science. For example, in biology and medicine, the importance of genes is often exaggerated by assuming that genes cause specific traits or diseases such as breast cancer. However, a gene by itself cannot do anything. What a gene does depends on its environment inside and outside the organism, as epigenetic biology emphasizes. In other words, **gene activity depends on its context** (Holdrege 1996, Hubbard and Wald 1999, Lipkin 2008, Dupré 2012). **Since the context also includes consciousness, thoughts, and emotions, they too may influence gene expression, which is relevant to sickness, health and healing and recognized in epigenetic medicine** (Church 2009). In all respects, "we are not determined by our genes, although surely we are influenced by them" (Lewontin 1991, p. 26). But "even if I knew the complete molecular specification of every gene in an organism, I could not predict what that organism would be...variations among individuals within species are a unique consequence of both genes and the developmental environment in a constant interaction" (ibid.). Unfortunately, the general public is often misled by scientists who, maybe through sloppy language, insinuate that genes determine healthy and pathological development.

Holistic Science - **Holistic science recognizes that the whole is more than the sum of its parts**. Thus, an organism is more than the sum of its components and its molecules. It has emergent properties. For example, a bird can fly, but its organs and its molecules cannot. A human being can think, but its DNA cannot. Thinking is an emergent property that is not found in the constituent parts. Furthermore, the parts can be understood only in their context within the whole. "The parts are not merely building blocks for the whole, but bear the impress of the whole throughout their nature" (Kirchoff 1995, p. 574). "In this mode of consciousness, we are able to see the whole as unitary and primary" (ibid.; see also Harman and Clark 1994, Harman and Sahtouris 1998, Goodwin 2001, 2007).

In general, holism emphasizes wholes and wholeness, as the name indicates. However, wholes may be seen differently. They may be seen just as material wholes. Hence, there is a holistic materialist science. For example, to a great extent, ecology appears more or less holistic in this materialist sense. But deep ecology goes beyond materialism. Ecosophy also transcends materialism as it integrates wisdom and love with ecology and economy (Sahtouris 2014).

To avoid the dualism of matter/energy and the non-material beyond matter/energy, Ken Wilber, following ancient traditions, distinguishes three levels of "matter" (or energy): the gross, subtle, and very subtle (causal) (Wilber 2000, 2006a,b). The gross level corresponds to the physical (matter/energy) as usually conceived in mainstream science. The subtle and very subtle (causal) levels represent **subtle and very subtle bodies or energies.** Holism, in a more profound sense, includes subtle and very subtle energies. As long as theories couched in terms of subtle and very subtle energies are testable, they can be considered scientific theories (see, for example, Tiller 1997, 2007, Manek 2019, Jabs and Rubik 2019). Some authors distinguish more than three levels of energy, and Greene (2009) emphasizes that they all form a continuum. This continuum comprises vital, emotional, mental, and universal energy (see Appendix 2). Greene also refers to "**inergy**," the combination of information and energy. Mainstream scientists, being imprisoned in the materialist worldview, reject the energy continuum as well as different levels of energy. This shows how scientific progress can be obstructed by dogma that is perpetuated by our mainstream culture and mainstream science.

If, however, the human body is understood in the most comprehensive sense, including the whole **inergy continuum**, it can become the doorway to the universe, universal consciousness or fundamental consciousness (Blackstone 2008). Being deeply aware of one's body creates an awareness of the continuum between one's body and its environment that extends into the universe. Thus, the body becomes transparent so that the commonly perceived illusory separation between the body and its environment disappears. Hence, the awareness of the oneness with the universe so that one can say "I am the universe" (Chopra and Kafatos 2017). This awareness cannot be gained if the body is seen only as a physical mechanism according to the prevalent mechanistic materialist science.

In his book *The Science Delusion* (2012/20), Rupert Sheldrake shows how science has been unnecessarily limited by assumptions that have hardened into dogmas. Sheldrake shows that **the worldview of materialistic mainstream science has become a belief system that has severe limitations**. For example, trillions of dollars have been spent on cancer research, and besides some limited progress, no general cure has been found (see, for example, Raza 2019).

In the skeptical spirit of open-minded scientific enquiry, Sheldrake turns ten fundamental dogmas of mainstream science into questions for open-ended research and provides evidence that supports the following **alternatives to the mainstream dogmas**:

1. Organisms cannot be fully understood in terms of the prevailing mechanistic materialist worldview.
2. Consciousness may be all-pervasive.
3. The total amount of matter and energy may change even after the Big Bang.
4. The so-called laws of nature may be only habits that can change and evolve.
5. Nature may be endowed with purpose.
6. Children may inherit characteristics acquired by their parents.
7. Minds may extend far beyond brains.
8. Memories may not be stored as traces in our brains and therefore may not be wiped out at death.
9. Psychic phenomena such as telepathy need not be discounted.
10. Mechanistic mainstream medicine can be complemented by alternative holistic approaches. Needless to say, these ten alternatives should not be understood as new dogmas but as alternatives for open-ended inquiry. If necessary, they should be modified or even abandoned as new evidence accumulates.

Dr. Larry Dossey, author of *Reinventing Medicine, One Mind* and many other books, wrote the following about Sheldrake's book: "Rupert Sheldrake may be to the twenty-first century what Charles Darwin was to the nineteenth: someone who sent science spinning in wonderfully new and fertile directions." And Dr. Andrew Weil, author of *Health and Healing: The Philosophy of Integrative Medicine* and many other books, wrote: "This provocative and engaging book will make you question basic assumptions of Western science." With regard to medicine, he added: "I agree with Rupert Sheldrake that, among other problems, those assumptions [of mechanistic materialism] hinder medical progress because they severely limit our understanding of health and illness."

One should, of course, not conclude that the mechanistic materialist approach is all wrong and Sheldrake's ten alternatives are right. Mechanistic materialism has made many important contributions to our understanding of reality, and Sheldrake's alternatives remain open to discussion. Openness is required so that science can advance. But the widespread mechanistic materialist dogma hinders progress and impoverishes our lives and society. "Reality, including our own existence, is so much more complex than objective scientific materialism allows" (Dalai Lama 2006, p. 39).

The Galileo Commission Project – It is reported that Galileo's contemporaries who opposed him, refused to look through his telescope; they refused to consider evidence that contradicted the prevalent dogma. Their refusal has striking parallels today.

Mainstream scientists and the general public who embrace mechanistic materialism refuse to consider evidence that contradicts their materialist dogma. To draw attention to this medieval attitude, the Scientific and Medical Network (www.scimed.org) established the Galileo Commission, represented by many advisers affiliated to thirty universities worldwide. Professor Harald Walach was commissioned to write a report entitled "Beyond a Materialist Worldview. Towards an Expanded Science." This report, which is available on the Internet (www.galileocommission.org), comprises a summary of fourteen arguments in favour of an expanded science and lists some of the **evidence that contradicts the materialist worldview:**

1. Near-death experiences (NDEs) in the absence of brain activity,
2. Non-local perception, also in the absence of brain activity,
3. A large database of research in parapsychology supported by meta-analyses,
4. A large database of children who remember previous lives (see also Walach 2018, Alexander 2018, Gober 2018a,b).

Nonetheless, materialists keep insisting that none of these phenomena exist or have not been sufficiently demonstrated. In this insistence, they "have to deny, erase, and take off the table so much of human experience to retain the illusion of the completeness of the materialist model" (Kripal 2019, p. 125). They have elevated this model to an ideology. Thus, we have gone almost full circle: during the Renaissance and subsequently during the scientific revolution, science liberated society to a great extent from Church dogmatism that repressed open enquiry, and today dogmatic materialist mainstream science debunks and represses holistic science and to a great extent keeps society in the dark. But although materialist mainstream science has led us into another dark age, there may be hope that eventually holistic science may succeed in a new liberation if it can surmount the enormous repressive power of mainstream science that has become official government policy in most countries.

Beauregard et al. (2014) issued a "**Manifesto for a Post-Materialist Science**" similar to the Galileo Commission Report. They concluded: "The shift from materialist science to post-materialist science may be of vital importance to the evolution of the human civilization. It may be even more pivotal than the transition from geocentrism to heliocentrism" (see also Grof 2006).

Consciousness - According to materialist science, consciousness is an epiphenomenon of the brain in which it resides. But how can the physical brain produce consciousness? This problem, which has been called the "hard problem," disappears if consciousness is considered primary or if consciousness and matter are seen as aspects of a deeper reality. From this perspective, consciousness is not produced in the brain. It exists beyond the physical body and brain, which then act as a filtering mechanism to produce individual consciousness (see, for example, Luisi 2009 (Chapter 6), Sheldrake 2012/20, Dalai Lama 2018, Alexander

2018, Gober 2018a,b, Hoffman 2018, 2019, Kastrup 2019, Kripal 2019, Presti et al. 2019, Schwartz et al. 2020). Some quantum physicists endorse this point of view. Max Planck, the founder of quantum physics, wrote: "I regard consciousness as fundamental. I regard matter as derivative from consciousness. We cannot get behind consciousness. Everything that we talk about, everything that we regard as existing, postulates consciousness" (quoted by Kripal 2019, p. 113). And the quantum physicist Erwin Schrödinger wrote: "consciousness is absolutely fundamental. It cannot be accounted for in terms of anything else" (quoted in Wikiquote; see also Spiro 2015). Some holistic biologists came to the same conclusion. Elisabet Sahtouris (2003) wrote: "consciousness is not a late emergent product of material evolution but the exact opposite, the source of all material evolution. Thus, **matter appears to be the contents of non-local infinite consciousness**; "the assumption that things exist as such, even outside and independently of our consciousness, is really absurd" (Schopenhauer, quoted by Kastrup 2020).

Individual consciousness arises in a more inclusive universal consciousness, what Judith Blackstone (2008), the Dalai Lama (2016) and others called fundamental consciousness: subtle luminous spaciousness (transparency) pervading your body and environment as a unity. Sages of the East and the West have been aware of this fundamental or universal consciousness for a long time. Jawer (2021) thinks that "sentience – the ability to feel – underlies consciousness," and therefore he considers sentience even more fundamental than consciousness. But note, he "thinks" so. Personally, instead of or in addition to fundamental or universal consciousness and sentience, I prefer to refer to the unnamable or mysterious.

> "Dualism in a Nutshell. What is Mind? No Matter. What is Body? Never Mind." (Cathcarts and Klein. 2010, p.102).

The Feminine Side of Science - In her book *Lifting the Veil: The Feminine Face of Science*, Shepherd (1993) explored the values, objectives, and results of science that includes the feminine and honours "the voices of women and men scientists as they reveal to us how the Feminine can make science more creative, more productive, more relevant, and more humane." (Shepherd 1993, p. 50). A common masculinist bias limits science unnecessarily, whereas the recognition of the feminine side of science extends its scope. The masculinist approach to science emphasizes cold analysis, competition, domination and often uses a language and metaphors of attack, war, and destruction such as, for example, the war against cancer. (Lakoff and Johnson 2003). In contrast, a more feminine approach to science emphasizes intuition, feeling, love, respect, nurture, interconnection, cooperation, etc. This emphasis changes the way science is done and its results. For example, in her investigations of the genetics of corn, Nobel laureate Barbara McClintock developed a "feeling for the organism" (Keller 1983) and "embodied values of the Feminine. As a geneticist, McClintock approached her object of study with reverence and humility. Rather than separate herself emotionally from her object of study, she became intimately involved with her corn plants. In describing her work, her vocabulary is one of affection, kinship and empathy, rather than that of battles, struggles, or a sense of opposition… For McClintock, science is not based on a division between subject and object, but rather on attentiveness as a form of love" (Shepherd 1993, p. 70). Such a loving relationship changes the interaction of the observer and the observed and therefore may reveal aspects that cannot be seen in an unloving interaction. This is especially evident in investigations of humans and animals. "Accepting an aesthetic based on love allows a different relationship to phenomena than is found in contemporary Western science" (Kirchoff 1995, p. 573). It can lead to "the reenchantment of science" (Griffin 1988, Sahtouris 2014).

Based on his doctrine of sociobiology in which the passing on of genes and competition plays a central role, E. O. Wilson concluded: "It pays males to be aggressive, hasty, fickle, and undiscriminating. In theory, it is more profitable for females to be coy, to hold back until they can identify males with the best genes…Human beings obey this principle faithfully" (quoted by Shepherd 1993, p. 48). However, "with careful observation of primate behaviour, feminist researchers are dispelling the myth of the active, courting, promiscuous male and the passive, coy, faithful female. Their research reveals ways in which females play an active role in sexual courtship. For example, Jane Goodall describes the prodigious activity of the chimpanzee Flo, who presented herself multiple times to all the males in the vicinity during estrus" (Shepherd 1993, p. 48). In general, Jane Goodall's non-intrusive and loving approach towards the chimpanzees she studied in their natural environment revealed aspects that could not be seen in caged chimpanzees, especially when they were treated unlovingly as mere research objects.

You should, of course, not conclude that "male science" is carried out only by male researchers and "feminine science" only by female scientists. At least some scientists, both male and female, may be able to balance more or less masculine and feminine approaches to science, which seems natural since, according to Daoism, Yang contains Yin and vice versa, and Yin and Yang form a continuum.

Broad Science - In his *Theory of Everything* (2001) and his earlier book *The Marriage of Sense and Soul* (1998), Ken Wilber contrasts narrow and broad science. Narrow science is based on sense data of the exterior world, whereas **broad science includes data of both the exterior and interior world**. Whether narrow or broad, Wilber suggests that science operates through the following three steps:
1. A practical injunction: "If you want to *know* this, you must *do* this – an experiment, an injunction, a pragmatic series of engagements, a social practice" (Wilber 2001, p. 75). For example, if you *want* to know whether Jupiter has moons, you

must look through a telescope. Or, if you *want* to know the effects of meditation, you must practice meditation. You cannot just rationally argue about it. You must practice it if you *want* to know it. Otherwise, refusing to practice it would be like refusing to look through the telescope.

2. Experience: "Once you perform the experiment or follow the injunction – once you pragmatically engage the world – then you will be introduced to a series of experiences or apprehensions that are *brought forth* by the injunction... you can have physical experiences (or physical data), mental experiences (or mental data), and spiritual experiences (or spiritual data)" (ibid., p. 75). In the above examples, you will experience moons around Jupiter (a physical experience), and you will experience the effects of practicing meditation (spiritual experiences).

3. Communal checking: "it helps if we can check these experiences with others who have also completed the injunction and seen the evidence" (ibid., p. 75). In this way, we obtain additional confirmation or disconfirmation of our insights, which, however, may not always be as straightforward as imagined by Wilber (see, for example, Lancaster 2004, p. 38).

An "extended science" that transcends the unnecessary limitations of ordinary mainstream science has also been proposed by other authors (see, for example, Josephson and Rubik 1992).

The general limitations of science (not the unnecessary limitations) that I discussed above (Chapters 1-7) also apply to extended science and broad science.

The Science of Enlightenment - Following the principles of broad science, spiritual practices can become at least in part scientific. "They rely on specific social *practices* or injunctions (such as contemplation); they rest their claims on data and experiential evidence; and they constantly refine and check these data in a community of the adequate [that is, those who have practiced the injunction(s)] – which is why they are correctly referred to as the contemplative sciences" (Wilber 2001, p. 77). In *The Science of Enlightenment*, Shinzen Young (2016) explains how the practice of concentration, clarity, and equanimity (non-interference) leads to insight and purification, which eventually may lead to enlightenment (see also Shinzen Young 2011-2016). As is often the case in the sciences, this prediction can be made only in terms of probabilities, not certainty. And it seems that the chances of reaching full enlightenment are rather low. It has been pointed out by other spiritual masters that concentration may lead to tension that may become a hindrance for reaching enlightenment (see, for example, Osho 2004, Bodian 2017).

In addition to showing a path towards enlightenment, science can reveal objective correlations between subjective meditative experience and enlightenment on the one hand and physiology on the other; in other words, correlations between subjective inner experience and objective scientific observation (see, for example, Goleman and Davidson 2017). It has been shown, for example, that meditative states are correlated with certain brainwaves or no brainwaves at all (Wilber 2006c). Such correlations can be investigated objectively and communicated through language. But the subjective experience of a deep meditative state and enlightenment is beyond language and thought, hence beyond science. Ken Wilber's AQAL Map clarifies this limitation of science.

The Big Three of Ken Wilber's AQAL Map - Ken Wilber's AQAL Map of human existence and the Kosmos has four quadrants or dimensions (Wilber 2000, 2001). (Wilber spells Kosmos with a capital K to indicate that it includes all dimensions of reality, not only the material cosmos). When the two right quadrants of his AQAL map are combined, we obtain **the Big Three: Science, Art, and Culture, or Nature, Self, and Morals.** This shows the limitations even of broad science because "science and its methods are still only "one third" of the total story" (Wilber 2001, p. 157). The other two thirds, the other two dimensions, are culture and art. Culture includes morals and often also religion, but the Dalai Lama (2012) has pointed out that ethics need not depend on religion. Art or the self entails subjectivity. Hence, subjectivity constitutes one of the three fundamental dimensions of human existence in the Kosmos. Yet in our scientific age, which values most highly objectivity, subjectivity is often devalued and relegated to an inferior status. This means devaluing one fundamental dimension of our humanity, which leads to an imbalance that entails a degree of insanity. Lived subjective experience transcends science. It may open the door to infinity beyond the illusion of a separate self. Recognizing this illusion leads to greater sanity (see Appendix 2).

Besides the big three dimensions, Wilber distinguished different levels that, for the self, range from divisive consciousness to unitary and nondual consciousness. In a simplified version, he referred to body, mind, and spirit, whereby mind comprises the body, and spirit includes both the body and the mind. This way, he avoided a dualism of body/mind and body/spirit (for limitations of Wilber and his AQAL Map see Sattler 2008, 2014).

Some authors have proposed a spiritual or sacred science that would include all dimensions of human experience (see, for example, Steiner 1986, Taylor 2018). However, Ravindra (2015-16) pointed out that spirituality and the sacred "is beyond all categories of interest available to science," and therefore he entitled his books more appropriately *Science and Spirit* (1991) and *Science and the Sacred* (2000), not "Spiritual Science" and "Sacred Science."

A Mandalic View - **I have indicated the difference and relation between materialism, holism, and what lies beyond in the form of a mandala, of which I presented a conceptual and pictorial version** (Sattler 2016b). The centre of the mandala represents the unnamable. It is surrounded by an inner circle that represents holism, including holistic science, and an outer circle that represents materialism, including materialist science. Holism, as the inner circle, is closer to unnamable reality than materialism, which means that it conveys a more encompassing view of reality than materialism. But even holism, including holistic science, constitutes only a map of reality, not reality itself that is beyond language, thought, and images. As a whole, the map of the mandala indicates a nondual vision of reality that comprises the unnamable, the mysterious source of everything, and the namable of which holism and materialism are represented. According to this vision, the unnamable and the namable are one (not two, nondual). Thus, the centre of the mandala can be seen as the source out of which holism and materialism arise.

The mandala can be extended and transformed in many ways. Therefore, it can be seen as a dynamic mandala of mandalas (see Sattler 2016b: Introduction and Chapter 7, and Sattler 2008: Chapters 4 and 5). Wilber's AQAL map can be seen as one of these transformations.

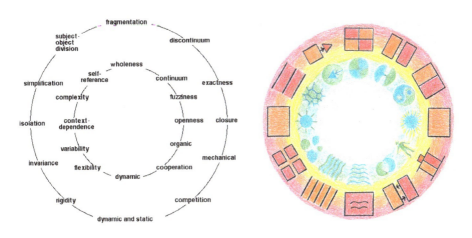

Conceptual version of the Mandala   Pictorial version of the Mandala

A MEDITATION

Experience yourself as light and thus transcend your illusionary separate self.
You might find it helpful facing the sun and thus breathing in the light and energy of the sun.
(for more on light meditations, see Osho 2010).

Or, if you find this meditation too difficult, you could do sky-gazing: Look at the blue sky, breath in the blue sky and then breath out into the blue sky to become one with it.

CONCLUSIONS

According to mechanistic materialist science, which still remains mainstream in our culture and science, reality such as an organism is seen as consisting of material components that interact as in a machine. In contrast, holistic science emphasizes wholes and wholeness. A whole is more than the sum of its parts. Therefore, even a material whole supersedes the mechanistic machine view. A more inclusive whole goes beyond materialism because it includes subtle energies and accepts phenomena such as extrasensory perception (ESP) that cannot be explained materialistically. Consciousness may be considered primary and universal.

Holistic science may also include the feminine side of science that emphasizes intuition, feeling, love, respect, nurture, interconnection, and cooperation. And holistic science may lead to broad science, which may facilitate a science of enlightenment. However, even broad science and a science of enlightenment represent only one of the big three dimensions of human existence: science, art, and culture or nature, self, and morals. Art and self may entail subjectivity. Yet in our scientific age, which values most highly objectivity, subjectivity is often devalued and relegated to an inferior status. This means devaluing one fundamental dimension of our humanity, which leads to an imbalance that entails a degree of insanity (see Appendix 2).

If the human body is understood in the most comprehensive sense, including the whole inergy continuum, it can become the doorway to the universe, universal consciousness or fundamental consciousness, which is beyond language and logic. I therefore refer to it as the unnamable or the mysterious beyond science. If we cannot see beyond science that relies on language and logic, our lives remain impoverished.

"If the universe means a vast machine to us, our whole being will unfold that meaning in the individual, in human relationships, and in society as a whole" (David Bohm).

"What we observe is not nature in itself but nature exposed to our method of questioning" (Werner Heisenberg).

"We construct our reality and then we construct mathematics to deal with our constructed reality" (Deepak Chopra).

"Matter is a human construct as is materialism" (Deepak Chopra).

"Reality, including our own existence, is so much more complex than objective scientific materialism allows" (the 14th Dalai Lama).

"The day science begins to study non-physical phenomena, it will make more progress in one decade than in all the previous centuries of its existence" (Nicola Tesla).

"We are like islands in the sea, separate on the surface but connected in the deep" (William James).

Chapter 9
Science, Society, and Culture

For a long time, Western society and its culture have been profoundly influenced by the Christian Church. But since society has become increasingly secular, science has replaced the religious influence to a great extent. Unfortunately, society is especially receptive to simple and sensational theories. "A simple and dramatic theory that explains everything makes good press, good radio, good TV, and best-selling books. On the other hand, if one's message is that things are complicated, uncertain, and messy, that no simple rule or force will explain the past and predict the future of human existence, there are rather fewer ways to get that message across. Measured claims about the complexity of life and our ignorance of its determinants are not show biz" (Lewontin 1991, p. VII).

Because of science's enormous impact on society, scientific advisors to governments should be well informed about the complexity and limitations of science, which, unfortunately, often is not the case. Insufficient knowledge does not serve the common good and the well-being of citizens, and prevalent misconceptions may have grave or even disastrous consequences.

As pointed out repeatedly, society is still dominated by mechanistic materialist science. What does not fit into this framework tends to be ignored or discounted as pseudoscience. The general public suffers from such shortsightedness and delusion. Healthcare is an example. As I pointed out in Chapters 2 and 7, the conservative medical establishment in collusion with the pharmaceutical industry and government uses its power to discredit and suppress alternative holistic medicine by all means available. The COVID-19 pandemic illustrates this very well.

The COVID-19 Pandemic - For governments and the World Health Organization (WHO) that were advised by the powerful medical establishment, the virus was the cause of the COVID-19 pandemic. This view may lead to **virus mania** if it is not recognized that, like a gene, a virus by itself cannot cause anything (Chapter 8). Therefore, as we have an epigenetic medicine, we also need epiviral medicine. Epiviral means 'over,' 'around' or 'beyond' the virus. Whether a person who contracts the virus gets sick depends on the person and the environment that are around or beyond the virus. If the virus *alone* were the cause of sickness and death, then everybody who contracts the virus should succumb to it. But this is clearly not the case. The virus appears to be a necessary but not a sufficient condition for the disease. **The disease is determined**

by many factors. Only one of these factors is the virus. But if the virus is considered *the* cause, then the focus is primarily on the virus. Therefore, declare war on the virus! And if it cannot be eradicated, then protection from the virus through hygiene (like handwashing), masking, and physical distancing is *the* solution. This view of the pandemic led many governments to draconian measures such as a widespread lockdown of society, which engendered enormous suffering, ruined the lives of many people and the economy. Bhakdi (2020a), a prominent microbiologist, referring to the German government, found such measures "grotesque, absurd, and very dangerous" because of the very destructive socio-economic consequences. Yet, most governments insisted that these measures were necessary, although Bhakdi (2020b), Reiss and Bhakdi (2020, 2021) and other scientists showed that they were not based on sufficient scientific evidence.

Eisenstein warned: "The measures being instituted to control Covid-19, likewise, may end up causing more suffering and death than they prevent… added deaths that might come from isolation-induced depression, for instance, or the despair caused by unemployment, or the lowered immunity and deterioration in health that **chronic fear** can cause. Loneliness and lack of social contact has been shown to increase inflammation, depression, and dementia. According to Lissa Rankin, M.D., air pollution increases risk of dying by 6 %, obesity by 23 %, alcohol abuse by 37%, and loneliness by 45%" (Eisenstein 2020a).

Nonetheless, prevention and treatment of COVID-19 were, of course, necessary. But in contrast to the narrowly restricted approach of the powerful medical establishment, **holistic medicine** offers a much more inclusive and successful approach in which, besides the virus, people and their environment must be considered. Since healthy people were not much affected by the virus, if at all, it appears crucial to support the physical, emotional, mental, and spiritual health of people, which can be done in many ways through healthy food, supplements (if necessary), and a healthy lifestyle, including exercise, relaxation and meditation (see Appendix 2). People in poor health who succumbed to the disease could be treated successfully by methods of alternative and integrative medicine unless they were already severely afflicted by other diseases. During the early stages of infection, before the virus reaches the lungs, humming can be helpful because it produces increased nitric oxide that has an antiviral effect, specifically inhibiting the replication of the virus in the nose and throat (Klinghardt 2020). In addition, nitric oxide can be administered orally. Cheng et al. (2020) considered large intravenous doses of vitamin C as the treatment of choice. Klinghardt (2020) found herbs such as *Andrographis paniculata* and synthetic products such as HOCl and chloroquine phosphate effective. In early stages of infection, Risch (2020), professor of epidemiology at Yale School of Public Health, treated his patients successfully with hydroxychloroquine (see also Mercola 2021). Risch concluded that "tens of thousands of patients with COVID-19 are dying unnecessarily" because hydroxychloroquine was not used. It was even censored because of potential side effects and for some patients it is counterindicated. But many drugs have side effects and are counterindicated for some patients. The MATH+

Protocol has been shown to be nearly 100% effective (for references, see Mercola 2020b). Yet this protocol and other successful alternative treatments were ignored by the powerful medical establishment, the World Health Organization (WHO)) and most governments. Why? Because, although they are based on compelling evidence, their safety and efficacy have not been proven. Of course, they have not been proven, because nothing, including conventional medicine, has or will be proven. As pointed out in Chapter 2, proof cannot be attained in science. But the powerful medical establishment misled governments with the myth that proof is necessary, and then, in turn, governments misled their citizens, and as a consequence, the harm and mortality inflicted were enormous because successful treatments were ignored and suppressed. This shows how a basic misconception about science can have devastating consequences. But they could have been avoided. We had solutions to the pandemic from early on in 2020. As Risch put it in July 2020: "The key to defeating COVID-19 already exists." But governments were blinded by the orthodoxy of the powerful medical establishment that refused to accept integrative medicine despite its successes. Integrative medicine accepts evidence-based methods in both alternative and conventional medicine (see Appendix 2).

Integrative medicine can also help to prevent infection. Lifestyle and food that support the immune system play an important role in prevention. Vitamins such as vitamin C and vitamin D have been shown to be effective. Further studies are underway (see Mercola 2020c,f). Humming can be useful because, as I pointed out above, it produces antiviral nitric oxide. Various herbal medicines and other methods may also be helpful. (Klinghardt 2020).

Except for very few governments such as China's, which recommended traditional Chinese medicine, most governments in collusion with the conservative medical establishment and the pharmaceutical industry failed to recommend evidence-based methods of integrative medicine that could have been helpful in the prevention and treatment of COVID-19. They let down their citizens. They did not even recommend food and supplements such as vitamins that boost the immune system. Their recommendations have not much advanced beyond the methods used over a hundred years ago during the Spanish Flu epidemic (1918-20). Like then, besides hygiene, physical distancing, and face masks, not much else was recommended. Holistic health care was ignored or discredited as non-scientific, pseudo-science, or conspiracy theory by not only most governments and the conservative medical establishment but also the WHO, which appears to be under the influence of the powerful medical establishment, the pharmaceutical industry, the Gates Foundation, etc. Evidence-based findings that questioned or contradicted the conservative agenda were censored. "YouTube's CEO even went on the record stating they will **block any video that goes against WHO guidance** on the COVID-19 pandemic. Shockingly, this includes videos and articles by medical doctors, emergency care specialists, and researchers who share their clinical experiences, recommendations, and scientific findings. Just one of countless examples was a May 17, 2020, Full Measure News report[1] in which Sharyl Attkisson interviewed doctors reporting good results with hydroxychloroquine. The segment also looked at the potential financial motives driving the mass media's disdain for the drug, while promoting remdesivir" (Mercola 2020e). "Open debate is stifled in favour of officially sanctioned orthodoxy" (Lorimer 2020), the orthodoxy that exercises power-knowledge (see Chapter 7). Unfortunately, with few exceptions, the media did not report evidence that contradicts the official narrative because such evidence is often dismissed as conspiracy theory (Davis 2020 and Chapter 7). Using the conspiracy label has become a weapon to dismiss and suppress evidence that questions or contradicts power-knowledge. Thus, the discussion of evidence is avoided. Criticism is silenced (Davis 2020).

In a broader general context, "another danger that is off the ledger is the deterioration in immunity caused by excessive hygiene and distancing. It is not only social contact that is necessary for health, it is also contact with the microbial world. Generally speaking, microbes are not our enemies, they are our allies in health. A diverse gut biome, comprising bacteria, viruses, yeasts, and other organisms, is essential for a well-functioning immune system, and its diversity is maintained through contact with other people and with the world of life. Excessive hand-washing, overuse of antibiotics, aseptic cleanliness, and lack of human contact might do more harm than good [Stromberg 2015]. The resulting allergies and autoimmune disorders might be worse than the infectious disease they replace. Socially and biologically, health comes from community. Life does not thrive in isolation" (Eisenstein 2020a; see also Bush 2020).

Therefore, **holistic health care recognizes the importance of the environment**. Water and air play an important role. "The national study from five Harvard researchers with the university's Department of Biostatistics analyzing more than 3,000 U.S. counties found that coronavirus patients with long-term exposure to fine particulate matter have significantly higher death rates than do those not subject to air pollution" (Bowden 2020). It seems that the virus can attach itself to fine particles in polluted air and thus can be transmitted through the polluted air (Gerretsen 2020). This may explain why some areas with much air pollution, such as Northern Italy, had a higher COVID-19 mortality rate.

According to some authors, electromagnetic radiation such as 5G technology has been related to COVID-19 (Firstenberg 2020, Rubik 2021). This contention remains controversial. There is, however, evidence that 5G and other electromagnetic radiation can be harmful to our health (Pall 2018, Mercola 2020a). Some people may be much more affected than others, which explains, at least to some extent, why this issue remains controversial.

The COVID-19 pandemic underlines the importance of a healthy environment and a sane society (see Appendix 2). Thus, it "creates a new opportunity to make a paradigm shift from the mechanistic, industrial age of separation, domination, greed and disease, to the age of Gaia, of a planetary civilisation based on planetary consciousness that we are one earth family. That our health is One Health rooted in ecological interconnectedness, diversity, regeneration, harmony" (Shiva 2020). Such health is not possible in rampant capitalism that exploits and destroys our natural environment. If COVID-19 originated through transmission from bats, it may be linked to the destruction of the forests in which bats live, so that, subsequently, bats moved into human settlements and then infected humans (for more examples, see Shiva 2020). If, however, SARS-CoV-2 was manufactured in a laboratory, it shows the danger of science and technology and its irresponsible use (for references, see Mercola 2020d). The "gain of function" research that produces more deadly viruses should be prohibited. As Lorimer (2020b) put it, we don't need gain of function, we need gain of wisdom (see also Bodian 2020).

In a critical evaluation of some of the major issues of the pandemic, Katz (2020) concluded: "There are real questions that should be addressed about dangerous biotech research, vaccine safety, patents and Pharma profits, restrictions on health freedom, and authoritarian tendencies in politics, finance, and technology." Restrictions on health freedom resulted from the power-knowledge exercised by the powerful medical establishment, the WHO, and governments, that is, the **medical-political complex** that includes the pharmaceutical industry. These restrictions led to much unnecessary suffering, mortality, and economic ruin.

To avoid misunderstandings, I have to add a comment about my references to Dr. Mercola's website (2020a-f). Although Dr. Mercola is a licensed osteopathic physician, in Wikipedia he is described as a charlatan. He favours alternative medicine, sells products of alternative medicine, but he also accepts evidence-based results of conventional medicine. In other words, he is a practitioner of integrative medicine. I referred to him repeatedly not so much because of his personal opinions but because he provides many references that include not only alternative medicine but also conventional medicine and research by critical medical experts, microbiologists, and epidemiologists at some of the best universities in America such as Harvard, Yale, and Stanford. In the mainstream media, the results of this research have been ignored, suppressed and even censured because it contradicted the power-knowledge of the powerful medical establishment. But as Robert F. Kennedy, Jr. emphasized: "The way you get democracy to function is by informing the public" (childrenshealthdefense.org).

Finally, we also have to see virus mania and the death toll of COVID-19 from a wider perspective. "The WHO reports that about 1.5 million people die every year from "diarrheal diseases"… an average regular winter flu kills about 400,000 of us every year… Road injuries are in the same range as diarrhea, or tuberculosis, with another 1.5 million gone with each, with Malaria running at about half that rate. That's nothing compared with the "4.2 million deaths every year as a result of exposure to ambient (outdoor) air pollution" or the "3.8 million deaths every year as a result of household exposure to smoke from dirty cookstoves and fuels" that the WHO reports. The World Bank calculates that this pollution drains the global economy of $5 trillion annually…The WHO also reports, "Recent evidence suggests that 134 million adverse events occur each year due to unsafe care in hospitals in low- and middle-income countries (LMICs), resulting in 2.6 million deaths annually"…if we are so disturbed by this epidemic and the number of humans dying, how come we allow some 15,000 children to die *every day* – that is about 5.5 million a year – from easily preventable diseases? How come we vote for politicians who supply killing machines to different parts of the world?" (OshoTimes 2020b).

Dr. Barke (2020) wrote: "In 1968, the Hong Kong Flu killed approximately 4 million people globally. Not only did we not shut down our economy in the face of that staggering number, but the three-day Woodstock Rock Festival in upstate New York was held in the midst of the epidemic. Isn't it time we stopped living our lives and dictating what we can and cannot do based on fear of every new danger?... Fear is far deadlier and a more contagious disease than COVID-19. Fear raises our blood pressure to unhealthful levels; fear influences us to make poor decisions. We fear being criticized, we fear exposing our ideas, we fear offending others, and we fear being infected by a virus that is not much more deadly than viruses of the past. As a physician with more than 25 years in practice, I will tell you what I tell my patients: you should not fear COVID-19. You should properly prepare and protect the most vulnerable in your homes, your businesses, and society, but you should join everyone else in living your life in maximum liberty with commonsense protections and precautions. I can also report that we have an effective treatment when symptoms of COVID-19 are seen early and remain mild. These treatments can also offer protection for the most vulnerable."

Realizing that the virus is not the cause of the disease but only one factor makes a huge difference. It empowers people and helps to reduce or eliminate their fear as they become aware of other factors that help to prevent infection and reduce or eliminate symptoms of the disease. **To a great extent, the confusion of cause and factor, that is, mistaking a factor (the virus) for the cause of the disease, has been at the root of the widespread virus mania, which shows how a fundamental misconception about science can have devastating consequences. The strategies based on this misconception were imposed by a powerful elite exercising its power-knowledge in a quasi-totalitarian manner in collusion with the pharmaceutical industry and governments, dismissing and suppressing evidence-based alternatives that could have reduced suffering and mortality to a great extent.**

Returning to science in general: governments and, to a great extent, society are aware of and use only a fraction of the available scientific evidence, which leads to a greatly impoverished and distorted view of science. Science by itself is already limited in many ways, as I point out in this book, but then governments and society become even more limited because of a narrow, one-sided and distorted view of science. This view then becomes power-knowledge, which means that governments and their advisors such as the conservative medical establishment and the WHO use their power to enforce this one-sided knowledge and to suppress evidence that contradicts it by declaring it misinformation, disinformation, unscientific, pseudo-science, or conspiracy theory. Unfortunately, the media often supports the one-sidedness. In this way, society, including the educational system, which affects young people as they grow up, is conditioned. If and when they become scientists, they have already been conditioned by a distorted view of science.

Furthermore, **science reflects values of society**. "Despite its claim to be above society, science, like the church before it, is a supremely social institution, reflecting and reinforcing the dominant values and views of society at each historical epoch (Lewontin 1991, p. 9). "Scientists do not begin life as scientists, after all, but as social beings immersed in a family, a state, a productive structure, and they view nature through a lens that has been moulded by their social experience" (ibid., p. 3). As Giere (2006) emphasized, **science reflects the historical, cultural, and social context in which it operates.** Despite this limitation, it is often believed that science reveals a kind of universal truth, when in fact truth is unattainable for science. This misunderstanding is at the root of many problems and can have grave and even devastating consequences. An example is the virus mania I referred to above that plunged societies all over the world into an economic disaster and enormous personal suffering. I find it astounding how so many intelligent and educated people can subscribe to and be imprisoned in the myths of modern materialist science. Obviously, indoctrination through the educational system and society has been very profound. Nonetheless, some great scientists have been able to see through the filters of society, at least to some extent.

Another example that shows how society influences science is Darwin's theory of evolution by natural selection. Darwin saw the struggle for survival in his society, and then he made struggle the major driving force for evolutionary change. He was also influenced by economic theorists of his time. "In fact, Darwin's whole theory of evolution by natural selection bears an uncanny resemblance to the political economic theory of early capitalism as developed by the Scottish economists" (Lewontin 1991, p. 10). It does not seem surprising that, even today, as capitalism rules most of the world, Neo-Darwinism still forms the core of biological science. Only a few biologists try to extend Neo-Darwinism, questioning the basic assumption that natural selection is the driving force of evolutionary change (see, for example, Wagner 2015). However, in the sixth edition of *On the Origin of Species* (1872), even Darwin admitted that "natural selection is not sufficient to explain the wealth of forms…in the various kingdoms of life" (Rutishauser 2019, p. 48). Furthermore, Darwin emphasized the importance of cooperation and love in animals and humans and a moral sense in humans (Loye 2018). But in our capitalist society, such comments have been largely forgotten because they don't resonate well with the capitalist agenda. Daniel Dennett, a well-known philosopher of science, called Darwin's theory of evolution through natural selection "the single best idea anyone has ever had" (Dennett 1995:21), but Lynn Margulis, a famous biologist who made fundamental contributions to our understanding of cell evolution, thought that history will ultimately judge Neo-Darwinism as "a minor twentieth-century religious sect within the sprawling religious persuasion of Anglo-Saxon biology" (Margulis, quoted by Mann 1991; see also Sahtouris 1999, 2000, 2003, Hands 2017, p. 575).

Much of society's ideological influence on science is less obvious and therefore more difficult to recognize. But it seems widespread. "It comes in the form of basic assumptions of which scientists themselves are usually not aware yet, which have a profound effect on the forms of explanations and which, in turn, serve to reinforce the social attitudes that gave rise to those assumptions in the first place" (Lewontin, p. 10). An example is the machine view of life. With the increasing importance of machines that led to the Industrial Revolution, living organisms were also seen as machines and this view of organisms is still firmly entrenched in the mechanistic materialist science of today. One consequence of this view is that, as a machine consists of separate components, an organism is also seen and studied as a conglomerate of parts. But this view misses the wholeness of the organism and the wholeness of nature (see Chapter 8). If we want to gain a deeper understanding of nature, we have to end the story of separation, which will end alienation and lead to a happier and more peaceful world (see Appendix 3).

Science and Morals – Morals appear embedded in the fabric of cultures. Are they related to science? It is often assumed that science and morals are separate domains. Thus, it is said that **science can tell us what *is* but not what *ought to be*. However, this distinction is not as clear-cut as is often assumed.** The science of animal behaviour has shown that "true pillars of morality, such as sympathy and intentional altruism, can be found in other animals" (de Waal 2005, p. 28). Hence, human morality has roots in our animal ancestry, which has been studied scientifically (see Appendix 3).

Kozlovsky (1974) emphasized the need for a naturalistic ethic guided by the common good for humanity and the environment. Ecological research has shown how human sustenance and survival are intimately linked to our environment. Hence, our values have to be informed by this research. In this sense, Sahtouris (1999, 2018) referred to an ecological ethics grounded in a deep understanding of the interconnectedness and balance of the natural world in which we are embedded. "We can -- and we must -- gain enough perspective to see ourselves as one part of a much greater living system, or being, and learn to act accordingly" (Sahtouris 1999). Thus, as nature can be a "source of guidance for human behavior, then surely science, as the study of nature, *should* concern itself with ethics -- with showing us what is wise or not wise to do in our relationship to one another and to the rest of the of nature" (ibid.). Such wisdom entails well-being. Thus, Harris (2010, p. 1) emphasized that, "questions about values...are really questions about the well-being of conscious creatures.

Values, therefore, translate into facts that can be scientifically understood." I think, however, that there remains a subjective aspect of values and well-being that is beyond science. It resides in the subjective experience of unity consciousness or nonduality, which can lead directly to moral behaviour. Unity consciousness and love or compassion "are one and the same" (Greene 2009, p. 214).

> In a Paris hotel elevator was written: Please leave your values at the front desk.
>
> The student said: I passed my ethics exam. Of course, I cheated.

Science and Religion – Religions as part of cultures may also influence science and vice versa. It is important to distinguish between religious experience and religious doctrine or belief. Religious experience or spiritual experience may be beyond discursive thought and language, and in this sense, transcend science. On the other hand, religious doctrine, dogma, and beliefs may be expressed through language, and then they may collide with science and even lead to an anti-scientific stance such as, for example, the refusal to accept biological evolution. But there are also religious doctrines compatible with science. For example, the Buddhist doctrine of impermanence agrees with science that emphasizes dynamics. But not all Buddhist doctrines agree with science. Therefore, the Dalai Lama, who valued scientific insights, affirmed that "some specific aspects of Buddhist thought – such as its old cosmological theories and rudimentary physics - will have to be modified in the light of new scientific insights" (Dalai Lama 2005b, p. 5). Wilber (2017, p. 78) asserted that "any spirituality that can't pass muster with science will not make it past the modern and postmodern tests for truth," but he also added that "any science that doesn't include some component of testable spirituality will never find an answer to the ultimate questions of human existence." The Dalai Lama also noted that "science will learn from an engagement with spirituality" (ibid.). But spirituality rooted in subjective spiritual experience goes beyond science (see Chapter 11). The same can be said for the religion that relies on religious experience, not dogmatic doctrine. Walach (2015) referred to "secular spirituality" as spiritual experience stripped of "doctrinal-dogmatic clothing."

"The true enemy of science is not religion. Religion comes in endless shapes and forms, and there are tons of faithful people with an open mind, who pick and choose only certain parts of their religion and have no issue with science whatsoever. The true enemy is …dogma" (de Waal, p. 109). Unfortunately, dogma exists not only in religion but also in science. But fortunately, not all scientists succumb to it.

The question of whether **God** exists does not appear to be a scientific question. It may be best approached from a first-person perspective, and then one has to be clear what is meant by *God*, since this word has many different meanings, ranging from one extreme, a father in heaven, to the another extreme, the sacredness or holiness of nature.

Since holiness is related to wholeness and since holistic science provides evidence for wholeness, a link may exist between holistic science, wholeness, and the holiness of nature. Being connected and one with this wholeness and holiness, we can feel "at home in the Universe" (Kauffman 1995) as the sacred (Kauffman 2010).

In his book *Beyond Religion* (2011), the Dalai Lama pointed out that **ethics is not necessarily dependent on religion.** He wrote: "I am confident that it is both possible and worthwhile to attempt a new secular approach to universal ethics…all of us, all human beings, are basically inclined or disposed toward what we perceive to be good… we have within our grasp a way, and a means, to ground inner values without contradicting any religion and yet, crucially, without depending on religion."

> "True religious mind is silent, free of fear and self-importance, innocent, open, and vulnerable, full of wisdom but unknowing, willing and able to be surprised" (Ravi Ravindra).

A Radical View of Science - **Paul Feyerabend** (1975a,b, 1978, 1987, 2011) found that the separation of science, society, religion, mythology, and the arts is not necessarily clear-cut. Consequently, he developed a view of science that is radical according to two meanings of the word "radical": 1. It appears extreme and shocking to most scientists, philosophers of science, and laypersons, and 2. It leads to the root meaning of science, which is knowledge. And knowledge can be obtained in many different ways: through observation, experiment, reasoning, feeling, emotion, intuition, and even magic and myth. Feyerabend excluded none of all that and demonstrated that practicing scientists may have recourse to all that and that the progress of science would not have been possible if all scientists had followed a rigidly defined scientific method. Therefore, he concluded that science cannot be defined by a so-called scientific method. Albert Einstein noted: "All science is nothing more than the refinement of everyday thinking." Since there are degrees of refinement, there seems to be a continuum from everyday thinking to scientific thinking, which means that science cannot be delimited from non-science. And, since, according to Feyerabend, no single methodological rule has been followed by all scientists, he claimed: 'anything goes' (as far as scientific methodology is concerned). I would prefer to say, 'many things go' (Sattler 1986, pp. 34/35 and 69).

In an article on the Shiva Nature of Science, Lane & Diem-Lane (2012) pointed out that, like Shiva with his many arms, "science is a quest with many methods and not just one," and therefore "science like Shiva cannot be confined to only one aspect… There are, in sum, innumerable ways to gather knowledge about the cosmos…Perhaps science's greatest contribution is that at its best it is open to refutation and is thereby open to change" (but see Chapter 2 about disproof and refutation).

However, Feyerabend noted that scientists often can be as dogmatic as people who subscribe to religious dogmas, myths, or ideologies (see also Lewontin 1991). For this reason, he considered science a form of religion or a pseudo-religion or ideology that has become repressive, although in the Renaissance, it started as a liberating movement. I would not say that all science is repressive, but the mechanistic materialist worldview of mainstream science has enslaved us. People who question it on the basis of good empirical evidence are no longer burned at the stake, but they are often ridiculed or threatened, and they may lose their research grants and even their positions. "Heretics in science are still made to suffer from the *most severe* sanctions this relatively tolerant civilization has to offer" (Feyerabend 1975b).

To protect citizens from the dogmas of science, Feyerabend thought that there should be a separation of science and the state, just as there is a separation of the church and the state. He said: "I want to defend society from all ideologies, science included. All ideologies must be seen in perspective. One must not take them too seriously. One must read them like fairy-tales which have lots of interesting things to say but which ... are deadly when followed to the letter" (Feyerabend 1975b). For example, the fashionable emphasis on neuroscience (brain research) can lead to a "neurocentric ideology" and "neuromania" (see Gabriel 2018, Bos 2017). And mainstream evolutionary theory (Neo-Darwinism) can become "Darwinitis" (Gabriel (2018). But, as I pointed out above, since Darwin recognized the limitations of his evolutionary theory, it is not quite fair to refer to Darwinitis. It would be more appropriate to call it Neo-Darwinitis since most defenders of Neo-Darwinism seem more dogmatic than Darwin.

Although I can see the merits of Feyerabend's radical view, I would not say that science *is* a dogmatic religion. I would prefer to say that science and dogmatic religion overlap in many instances. And I think that it is possible to be non-dogmatic in both science and religion. The problem is that most scientists do not even seem to be aware of their dogmatism because they have been deeply indoctrinated and brainwashed during their university education. Hence, the first task is to create awareness of the dogmatism. Then one may proceed in an open-minded and open-ended science...

Someone said: The mind is like a parachute: it works much better when it's open.

Although mainstream science remains predominantly materialistic and mechanistic, one can see in some instances that this dogmatism is giving way to a more inclusive holistic approach. For example, a new discipline called psycho-neuro-immunology is integrating aspects of the mind and body. Neuroplasticity shows that we need not be caught forever in rigid behaviours and views (Doidge 2015). And some medical doctors are now prescribing meditation in addition to or instead of drugs (see Appendix 2). Schroeder (2016) noted that "the Consortium of Academic Health Centers for Integrative Medicine has as its members' institutions with the highest reputation in the world, such as Johns Hopkins University School of Medicine or Mayo Clinic. It is a spectacular, victorious return of the idea that the primary goals of the discipline are wellness and healing of the entire (sic!) person in bio-psycho-socio-spiritual dimensions."

Nonetheless, as Ken Wilber pointed out, there is still a strong tendency to collapse the Big Three (art, culture, and science) into a Big One: science, empiricism, and objectivity, based on the belief that objective empirical knowledge of sensory experience is the only *true* knowledge. As a result, to a great extent, subjectivity has become taboo, as Wallace (2004) pointed out in his book, *The Taboo of Subjectivity*. Thus, life has become impoverished and unbalanced. To regain a healthy balance of the Big Three dimensions of human existence, we need to emphasize the importance of subjectivity of the self and art besides objectivity in science. Culture may include both subjectivity *and* objectivity.

Dangers of Science - Because of the limited insight of scientists, the consequences of scientific innovations often cannot be foreseen sufficiently. Thus, what appears beneficial at first may later turn out to be detrimental or even disastrous. And indeed, as Maxwell (2017) pointed out, science combined with technology has brought about "almost all our current grave global problems: rapid population growth, destruction of natural habitats and rapid extinction of species, the lethal character of modern war, the development of extreme inequalities of wealth and power around the globe, pollution of earth, sea, and air, and most serious of all, the impending disasters of climate change" (Maxwell 2017, quoted by Lorimer 2017). Many other dangers and risks could be added to this list, such as the risk of advancing artificial intelligence that, according to Stephen Hawking and others, may lead to the demise of humankind (see also Harari 2017). For this reason, we need the precautionary principle: let us not pursue risky scientific and technological research. Science is not sufficient to decide which scientific projects should be pursued and applied in technology. Such decisions may also involve moral, cultural, philosophical, and spiritual dimensions.

A CONTEMPLATION

Whatever happens, know that you are not this and not that because you are infinitely more than just this or that. Reside in the infinity beyond science out of which this and that arise.

CONCLUSIONS

Science and society interact and influence one another. Values of society may influence scientists because scientists grow up and are educated in a culture. But science and morals are often seen as distinct. It is often said that science investigates what *is* and cannot tell us what *ought to be*. However, if we agree that the highest value is the well-being of human beings and society, then science may be able to tell us, at least to some extent, what we should do to manifest this value. Nonetheless, some personal subjective decisions that transcend science may be unavoidable.

Science and religion may or may not be compatible. In the latter case, religious doctrines may have to be changed so that the conflict will be removed. But scientists may also learn from religious and spiritual insights. Religious or spiritual experience may go beyond science and religious doctrine in as much as it transcends language and logic.

Governments and a large segment of society often promulgate a simplified, one-sided, and distorted view of science. Thus, mainstream society is still dominated by mechanistic materialist science that disregards to a great extent holistic science and even rejects much of it. To a great extent, conventional medical science also still operates within the mechanistic materialist framework. Bio-psycho-socio-spiritual dimensions are not sufficiently recognized.

The **COVID-19 pandemic** exemplified how misconceptions of science and power-knowledge exercised by a powerful elite of the powerful medical establishment had devastating consequences that could have been avoided or reduced through the realization that:

1. the mechanistic materialist approach of the powerful medical establishment and the World Health Organization (WHO) was too limited to deal more successfully with the crisis;
2. vaccines, physical distancing, masking, and hygiene were not the only means to protect oneself from COVID-19;
3. the virus was not the cause of the disease but only a factor in the whole system that comprises people and their environment;
4. therefore, the focus should have been on people and their environment, not just on the virus, which can lead to virus mania;
5. alternative and integrative medicine had effective methods to prevent or reduce infection and to treat successfully people who succumbed to the disease;
6. these methods were ignored and suppressed by the collusion of powerful medical officers, the pharmaceutical industry, WHO, and most governments; they were considered unproven; but
7. all science, including conventional and holistic medicine, remains unproven (see Chapter 2). All we can have in science is evidence, not proof; and there is much evidence supporting holistic methods of integrative medicine;
8. the mainstream media failed to expose the misconceptions perpetuated by the powerful medical-political complex;
9. what was presented as "science" to the general public was only the one-sided power-knowledge of WHO, chief medical officers in powerful positions, the

powerful pharmaceutical industry and governments; much contradictory scientific evidence was ignored or suppressed;
10. from what we know, it seems that much personal suffering, much sickness and mortality, and much economic ruin could have been avoided if methods of integrative medicine would have been widely implemented instead of being ignored or suppressed as misinformation, disinformation, pseudoscience, or conspiracy theory.

As, in the past, we have had much persecution and destruction in the name of God, so now much unnecessary suffering is inflicted in the name of science, especially when it is equated with mechanistic materialist science and the still more restrictive view that a gene or virus is the cause of a disease, a view that is used as a weapon to suppress other scientific evidence, even evidence provided by medical doctors, microbiologists, epidemiologists, and public health experts who disagree with the power-knowledge of the official narrative. This shows how power can override scientific evidence.

Paul Feyerabend's radical view of science acknowledged science's interaction with society, religion, mythology, and the arts. Because of widespread dogmatism that narrows the scope of science, Feyerabend saw science as a form of religion. I think, however, that exceptional scientists may go beyond dogmatism and may see beyond the limitations of science. But dogmatism, whether in science or religion, constrains, becomes a prison, and then may lead to intolerance, antagonism, and even war. Rising beyond dogmatism opens doors to greater sanity, freedom, tolerance, and peace.

Despite the best intentions, scientists often may not be aware of the consequences of their research for society and the planet. What appears beneficial at first may later turn out to be detrimental or even disastrous. And indeed, as Maxwell (2017) pointed out, science combined with technology, has brought about "almost all our current grave global problems."

> "From narrowness to broadmindedness,
> From prejudice to tolerance.
> It is the voice of life that calls us
> To come and learn"
> (Bell inscription at the University of Buffalo)

Chapter 10
Science and the Arts

Because of its limitations, science can reveal only aspects of reality. The arts disclose other aspects. Thus, science and the arts complement one another. Arthur Koestler put it this way: "Einstein's space is no closer to reality than van Gogh's sky. The glory of science lies not in a truth more absolute than the truth of Bach or Tolstoy, but in the act of creation itself. The scientist's discoveries impose its own order on chaos, as the composer or painter imposes his, an order that always refers to limited aspects of reality, and is based on the observer's frame of reference, which differs from period to period as a Rembrandt nude differs from a nude by Manet" (Koestler 1964). But since the arts, especially the visual arts and music, go beyond language, they also transcend science, for which language remains fundamental.

Besides complementing one another, **science and the arts also overlap and may interact** (McLeish 2019). In their search for theories, scientists may be inspired by poetry, painting, music, and other arts. And artists may find an impulse through science.

Beauty is often associated with the arts, but it may also play a role in science. Although the aim of science is to search for theories and laws that provide explanations and predictions, **science has also revealed much beauty**. Think of the beauty of stars, nebulae, and galaxies, the beauty of plants and animals, and even the beauty of molecules such as the double helix of DNA. As a plant morphologist, I have been much impressed by the beauty of plant form.

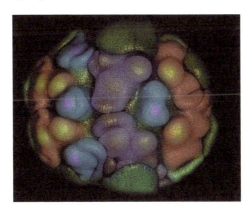

Photo of floral primordia of *Allium sativum* (garlic). Courtesy of Dr. Somayeh Naghiloo, who used a digital version of the author's epi-illumination technique.

Besides having *revealed* beauty, science has also *created* beauty. One of the most striking examples is fractal geometry. In contrast to traditional Euclidean geometry, which allows only simple structures such as circles, triangles, and squares, fractal geometry produces patterns of extraordinary complexity and beauty (see, for example, Peitgen & Richter 1986). To some extent, some scientific theories may also appeal to our aesthetic sensibility because of their elegance. Theories such as, for example, Einstein's theories of relativity, have an elegance that appears beautiful. Einstein occasionally valued elegance even more than facts, as when he said (maybe jokingly?): "If the facts don't fit the theory, change the facts." Besides elegance, simplicity may also be considered beautiful (Glynn 2013). Thus, the simplicity of trinities is often deeply appealing not only in religions but also in science. For example, according to the classical theory of plant morphology, there are only three organ categories: root, stem (caulome), and leaf (phyllome). Therefore, the whole diversity of plants, including flowers, is reduced to this trinity. The majority of plant morphologists still defend this trinity, although there is much evidence that it is a simplification that creates a distortion in many cases (see Appendix 1). Molecular biology also has a favourite trinity: DNA, RNA, and protein. According to the central dogma, information flows only from DNA to RNA to protein, but it has been shown that this dogma is not generally valid.

Some scientists see an intrinsic connection between beauty and scientific truth, which at best can be only a partial truth, as I have pointed out in preceding chapters, especially in Chapters 6 and 7. So these scientists say that, for a theory to be true, it must also be beautiful. In other words, if it is not beautiful, it cannot be true. Citing the poet John Keats, the quantum physicist Paul Dirac said: "beauty is truth, and truth is beauty." Thus, concerning Einstein's relativity theory, he concluded: "It is the essential beauty of the theory, which I feel is the real reason for believing in it" (quoted by McAllister 1999). With regard to the structure of DNA, Rosalind Franklin, who contributed to its elucidation, "accepted the fact that the structure was too pretty not to be true" (quoted by Nidamboor 2017). To what extent beauty and scientific truth coincide remains, however, debatable. In his book *Truth or Beauty*, Orrell (2012) presented several examples from the history of science that show how an ideal of beauty can be misleading for our understanding of the world. But beauty may play a heuristic role, which means that initially, it may direct the researcher toward beautiful hypotheses and theories, which then have to be validated by observations and/or experiments. It happens that the original beautiful hypotheses have to be given up when more complexities are revealed. For example, to explain the development of flowers an ABC model was originally proposed. But as research advanced, this beautiful trinitarian model had to give way first to an ABCD model and then to an ABCDE model.

Neither in science nor the arts should we make beauty an ideal or an obsession because "as soon as we hold on to anything its opposite arises: the obsession with beauty is in itself ugly" (Sabbadini 2013, p. 49). As Laozi reminded us long ago: "Recognize beauty and ugliness is born."

A MEDITATION

Listen to beautiful music or immerse yourself into a painting and become one with it, beyond language, thought, and science.

This meditation, like the others I suggested, may appear too ordinary to many people. But in deep listening, the unexpected may happen. Once, in a class, I turned off the lights and I asked the students to close their eyes. Then I played a piece of classical music and I asked them to immerse themselves in the sound. Afterward, several students told me that they had never before experienced music so deeply. One student said that it was the deepest experience of his life. Thus, the ordinary can become quite extraordinary if we can be totally present in the moment.

CONCLUSIONS

Science and the arts complement one another, but since the arts, especially the visual arts and music, go beyond language, they also transcend science, for which language remains fundamental.

Science and art overlap and may interact. Science has revealed beauty and has created beauty, such as the beauty of fractals. Theories can also be beautiful in their simplicity and elegance, but claiming that theories are only true if they are also beautiful appears exaggerated and questionable. Furthermore, scientific truth can only be an aspect of Truth, not Truth as that which is. Not recognizing this leads to degrees of insanity, that is, delusion about reality.

"I am among those who think that science has great beauty. A scientist in his laboratory is not only a technician: he is also a child placed before natural phenomena which impress him like a fairy tale" (Marie Curie).

"Science is not a heartless pursuit of objective information. It is a creative human activity, its geniuses acting more as artists than as information processors" (Stephan Jay Gould).

"All religions, arts, and sciences are branches of the same tree" (Albert Einstein).

"The aim of art is to represent not the outward appearance of things, but their inward significance" (Aristotle).

"Art must be an expression of love" (Marc Chagall).

Untitled (0-20-6) by Ulrich Panzer (2020). Acrylic ink on mylar, 42 x 42 inches. Courtesy of Ulrich Panzer.

Chapter 11
Science and Spirituality

I hesitate to refer to spirituality because it implies spirit, and spirit is often seen as opposed to matter, including our bodies. We can, however, overcome this dualism of spirit versus matter if spirit includes matter. According to Wilber (2000) and others, **spirit transcends and includes matter**. Alternatively, **one may refer to mystery, which supersedes the distinction of matter and spirit** (Ferrer 2017).

I understand *Science and Beyond* also in this nondualistic sense: what is beyond science does not exclude science; it transcends and includes science. Similarly, mystery is not opposed to science but includes it, or, as Albert Einstein wrote, science arises out of the mystery, which means that the mysterious is the source of science. Science, philosophy, art, and spirituality all arise out of this mysterious, unnamable source and thus present different aspects of reality.

It appears obvious that materialist science is not closely related to spirituality. However, it could be considered complementary to spirituality unless it takes materialism as the absolute truth, which seems a widespread conviction in the materialist scientism of our society. In that case, materialism functions like a dogmatic religion or pseudo-religion.

> "There is to be a christening party for Paddy and Maureen's new baby, but before the ceremony the priest takes Paddy aside and asks, "Are you prepared for this solemn event?"
> "I think so," replies the nervous Paddy. I've got cheese rolls, salad, and cake."
> "No, no," interrupts the priest, "I mean spiritually prepared?"
> "Well, I don't know, says Paddy thoughtfully, "Do you think two cases of whisky are enough?" (Osho 1998, p. 228).

In contrast to materialist science, holistic science is more closely related to spirituality and, to some extent, converges with spirituality. The convergence resides in the recognition of the interconnectedness of everything and oneness (see, for example, Capra 1975). As a result, many statements of holistic scientists sound remarkably similar to statements of spiritual masters and mystics who emphasize oneness and non-duality. McFarlane (2002) collected many such parallel statements in his book *Einstein and Buddha*.

Oneness is especially well-known in quantum physics, where it is recognized that the observer is integrated with the observed and that entangled particles that are propelled far away from one another remain entangled. And oneness has also been shown in other sciences (Hollick 2006, Sahtouris 1999, 2003, 2014, 2018). In plant biology, my field of specialization, it has become evident how plants are one with the universe because they are continuous with the soil and air around them; they are connected with the sun through solar radiation and with the universe through cosmic radiation. Ecology has shown the interconnectedness of everything. The recognition of interconnectedness leads to compassion.

Interconnectedness means that there are no separate things. This **no-thingness** has been much emphasized in Buddhism and other mystical teachings. Thus, no-thingness, interconnectedness, and non-separability form a bridge between mysticism, nondual spirituality, and holistic science.

Thich Nhat Hanh, the Vietnamese Zen master, referred to **inter-being**. For example, "you will see clearly that there is a cloud floating in this sheet of paper. Without a cloud, there will be no rain; without rain, the trees cannot grow; and without trees, we cannot make paper… so we can say that the cloud and the sheet of paper *inter-are*… Without "non-paper elements," like mind, logger, sunshine, and so on, there will be no paper. As thin as this sheet of paper is, it contains everything in the universe in it" (Thich Nhat Hanh 1988; see also Thich Nhat Hanh 2014). Similarly: "The whole cosmos has come together in order to help the flower manifest itself. The flower is full of everything except one thing: a separate self or a separate identity" (Thich Nhat Hanh 2003). The same can be said of humans: "The Hindu Upanishads… identify Atman, the ultimate nature of the self, with Brahman, the ultimate nature of the world. Neither exists independent of the other, and both are inseparable aspects of a single, nondual reality" (McFarlane 2002, p. 63).

Oneness or the One need not negate the Many. **Nonduality** recognizes that the One and the Many are not two (nondual), which can be understood as the most inclusive oneness (see, for example, Arber 1957). According to the Heart Sutra (of Mahayana Buddhism), form is emptiness, and emptiness is form. Form here means the Many, whereas emptiness implies no-thingness (no separate things). Emptiness is the translation of "sunyata" (Sanskrit) in the original text of the Heart Sutra. Tanahashi (2014) translated "sunyata" as boundlessness, which implies interconnectedness and oneness. Boundlessness or emptiness applies not only to the manifoldness of form but also to feelings, perceptions, mental formations, and consciousness (see Dalai Lama 2005).

Despite the astounding convergence of the deepest insights of holistic science and spirituality, we can also see a divergence: differences remain. One important difference is that holistic science, like all science, presents a third-person perspective of reality: holistic scientists talk about reality using language and the thinking mind. Spirituality may also be talked about, but central to it is the spiritual experience in which the spiritual master or mystic realizes his or her oneness with the universe. This experience or, rather, this state of being is beyond words. It cannot be talked about. And yet mystics talk about it, but they agree that words, even words like oneness, are inadequate to describe their experience, their beingness. Religious experience may also transcend words. It may coincide with spiritual experience inasmuch as it is at the root of the mystical traditions in all religions and thus it may transcend the doctrinal differences between religions. Much antagonism between religions could have been avoided if the root meaning of the word "religion" had been kept in mind, namely, that the word is derived from the Latin "religio," which means "to connect."

What lies beyond language can be experienced in **silence**. Hence, the importance of silence. It may open the door to the unnamable, which we may also refer to as the **mystery** (Wolfe 2014). Mystery here is understood as the ultimate mystery that is beyond knowing through the mind; hence, beyond the ego. "Spiritual contemplatives have had at least momentary experiences of becoming one with Ultimate Mystery" (Wolfe 2014, p. 33). Skeptics point out that such experiences may create only a feeling of oneness, but that this feeling is not founded in reality; in other words, that it is illusory. However, I think that the skeptics, not having had such an experience, cannot imagine the profundity of this experience and beingness. And they seem to ignore that this experience and beingness are corroborated by holistic sciences such as ecology and quantum physics, which have provided much evidence for interconnectedness and oneness.

It is reported that one day, as usual, a crowd had gathered to listen to the Buddha. But he did not speak; he only held a flower in his hand. The crowd became rather restless as he kept looking at the flower in silence. Finally, Mahakashyap, one of his disciples, started laughing. Then Buddha gave him the flower and said to the crowd: "Whatever can be said through words, I have said to you, and that which cannot be said through words, I give to Mahakashyap. The key cannot be communicated verbally...What is the key? Silence and laughter is the key - silence within, laughter without. And when laughter comes out of silence, it is not of this world, it is divine" (Osho 2004, pp. 202-203). Mahakashyap has been considered the forerunner of Zen Buddhism that emphasizes the limitations of language and thought.

John Cage made history with his "composition" 4'33," which originally, in 1952, was performed by a pianist sitting in silence at the piano for four minutes and thirty-three seconds. Silence also played a role in some of Cage's other compositions and his life, which was deeply inspired by Zen Buddhism. In our society, silence is often misunderstood and not appreciated. Many people feel embarrassed when they are together in silence. They feel compelled to talk, not knowing that the deepest communication may happen in silence. But it cannot be imposed. It has to happen spontaneously. Imposed silence seems like repressed noise. It is not genuine. Furthermore, as Cage pointed out, silence cannot be separated from sound. Even in a completely soundproof chamber, we can still hear sounds, such as the sound of our heartbeat. Hence, in a nondual sense, silence and sound are one, not two. Thus, we can be silent even in very noisy places, even in the marketplace.

In a nondual sense, the speakable and the unspeakable (to which Korzybski referred to) or the namable and the unnamable are one (not-two). The namable and the mystery of the unnamable arise from a yet deeper mystery that is referred to in the *DaodeJing* (*Tao Te Ching*) as "the darkest of the dark, the door of all mysteries" (Sabbadini's 2013 translation) or "darkening that darkness, that is the door of all wonders" (one of the alternative translations included by Sabbadini 2013).

Meditation may be helpful to enter "the door of all wonders." To many people, meditation means only sitting meditation. There are, however, many different ways of meditation, including dynamic meditation. The ancient Vigyan Bhairav Tantra describes 112 meditation techniques (see Osho 2010). Among these techniques, you may find one that is particularly suited for you. Osho (2004) devised meditation techniques that may be especially useful for modern people who find it difficult to relax. Techniques function only as doors into meditation beyond technique. Osho said that if meditation succeeds, no religion will be needed, no god. However, for some people who suffer from deep-rooted psychological problems, meditation may be inappropriate or even dangerous (Grof and Grof 1989, OshoTimes 2020). Music, singing, chanting, dancing, and laughing may also be doors into the infinite mystery. In total dancing, the dancer may become the dance in which the separate ego is transcended. Total laughter has been described as a shortcut to nirvana (liberation) because the thinking mind is transcended. One cannot laugh totally and think at the same time. As one transcends the thinking mind, one also goes beyond science, which functions with the thinking mind.

The arts, especially visual arts and music, may also become a door to mystery. Thus, they are not only complementary to science, but they may also lead us beyond science. Furthermore, as the renowned dancer Ruth St. Denis said, "suddenly, in any mundane moment, the infinite may come through." Thus, everything can become a meditation or a new way of being.

In the widest sense, we can envisage that "human spirituality is "participatory" in the sense that is can emerge from (a) the active participation of all human dimensions (body, instincts, heart, mind, and consciousness); and (b) our co-creative interaction with a dynamic and indeterminate power or Mystery" (Ferrer 2004; see also Ferrer 2017).

All these experiences of mystery are gained through deep personal experience and insight and therefore entail **subjectivity**. They express the first-person experience, whereas science implies third-person experience. They transcend language and the associated thinking mind, whereas science functions with language and the thinking mind. Although subjectivity may lead us astray, it has the potential to reveal the deepest insights, insights that cannot be gained through science because of its limitations. In our age, which is dominated by science and its ideal of objectivity, I consider it important to draw attention to subjectivity, to personal insight beyond science. Let us not forget that some of the deepest insights of humankind were gained through personal subjective experience, insights like those of the Laozi (Lao Tzu) in the Daodejing (Tao Te Ching) (p. 88), the Buddha (p. 79), Heraclitus (p. 92), Attar of Nishapur (p. 81), Rumi (p. 129), Gurdjieff (p. 127), Nisargadatta (p. 128), Osho (pp. 14, 79) Thich Nhat Hanh (pp. 78, 131), the 14th Dalai Lama (pp. X, 81-82), and many other sages of the East and West. Intuition and personal insight also play a role in science, but they are subjugated to rational inquiry and empirical evidence and expressed through language. Not recognizing this leads to degrees of **insanity**, delusion about reality.

In the long poem, *The Speech of Birds* or *The Conference of Birds* (1177), by the Sufi poet Farid ud-Din Attar, commonly known as Attar of Nishapur, one bird, the wisest of them, leads the others through seven valleys towards the mysterious Simorgh, their king, who symbolizes enlightenment. In Wikipedia, these seven valleys are described as follows:

1. Valley of the Quest, where the Wayfarer begins by casting aside all dogma, belief, and unbelief.
2. Valley of Love, where reason is abandoned for the sake of love.
3. Valley of Knowledge, where worldly knowledge becomes utterly useless.
4. Valley of Detachment, where all desires and attachments to the world are given up. Here, what is assumed to be "reality" vanishes.
5. Valley of Unity, where the Wayfarer realizes that everything is connected and that the Beloved is beyond everything, including harmony, multiplicity, and eternity.
6. Valley of Wonderment, where, entranced by the beauty of the Beloved, the Wayfarer becomes perplexed and, steeped in awe, finds that he or she has never known or understood anything.
7. Valley of Poverty and Annihilation, where the self disappears into the universe and the Wayfarer becomes timeless, existing in both the past and the future.

Science may traverse some of these valleys at least to some extent, but since all scientific knowledge is expressed through language that uses logic, science remains limited. Hence, it cannot quite reach Simorgh (that is beyond words and logic), but maybe it can point in its direction (see, for example, Sattler 2016). Science cannot spare us the often-arduous journey through the valleys, in which many may get trapped and even perish (like many of the birds that symbolize humans). The few birds (humans) that reach Simorgh learn that they themselves are Simorgh. In other words, we are what we are searching, as the nondual masters like Laozi (Lao Tzu) have told us.

Although we have to go beyond language and thought on our personal subjective journey, let us not turn against the thinking mind that is a basic tool for scientific inquiry and our orientation in the world. Let us integrate head, heart, and gut. As proposed in Wilber's integral AQAL Map of the Kosmos and human existence, let us integrate nature, culture, and the self, or science, morals, and art. And let us not forget that even the best map is not the territory of reality, which is beyond maps, beyond language and thought, beyond science, and even beyond the beyond in as much as the beyond is communicated through language and logic. Thus, as expressed in the mantra of the Heart Sutra: "Go, go, go beyond, go totally beyond, be rooted in the ground of enlightenment" (translation by the Dalai Lama 2005).

A Zen student asked his master: 'Is it okay to use email?' 'Yes,' came the reply, 'but no attachments!'

A MEDITATION

With eyes closed, inhale and then, as you exhale, chant OM, or listen to a fading sound, such as the sound of a bell. Thus, dissolve in infinity, beyond language, thought, and science.
(for more on this meditation, see Osho 2010 p. 380)

CONCLUSIONS

In contrast to materialist science, holistic science is more closely related to spirituality and, to some extent, converges with spirituality. The convergence resides in the recognition of oneness in holistic science and spirituality. However, differences remain between the two. One important difference is that holistic science, like all science, presents a third-person perspective of reality: holistic scientists talk about reality using language and the thinking mind. Spirituality may also be talked about, but central to it is the spiritual experience in which the spiritual master or mystic realizes his or her oneness with the universe. This experience, or rather this state of being, is beyond words. Religious experience may also transcend words. Thus, religious experience and spiritual experience may coincide or may be just different words for the same or similar experience. It seems that religious experience is at the root of the mystical traditions in all religions and transcends the doctrinal differences between religions.

What lies beyond language can be experienced in silence. Hence, the importance of silence. It may open the door to the unnamable, which we may also refer to as mystery. Mystery here is understood as the ultimate mystery that is beyond knowing through language, the thinking mind, and science, hence, beyond the ego. Although beyond, it includes all. Thus, the mysterious can be understood as the source of science, philosophy, art, and spirituality.

Meditation may be helpful to go beyond language and the thinking mind. Furthermore, such transcendence may happen through art, music, singing, chanting, dancing, laughing, or just spontaneously. In any case, it expresses a personal, subjective experience, a first-person experience, whereas science implies third-person experience. Although subjectivity may lead us astray, it may also lead to the deepest insights beyond science, like those of the Laozi (Lao Tzu), the Buddha, Heraclitus, and many other sages of the East and West.

With the rise of science and its emphasis on objectivity, the importance and relevance of subjectivity have been questioned and negated to a great extent. As a result, our personal lives and society have become impoverished and unbalanced. To regain a healthy balance, we need the recognition of subjectivity along with the objectivity of science. And subjectivity needs to be deepened through greater awareness of the unnamable mystery.

> "The most beautiful and deepest experience a man [and woman] can have is the sense of the mysterious. It is the underlying principle of religion as well as of all serious endeavour in art and science…To sense that behind anything that can be experienced there is a something that our mind cannot grasp and whose beauty and sublimity reaches us only indirectly and as a feeble reflection, this is religiousness. In this sense I am religious" (Albert Einstein).

> "Silence is the best language. More is communicated through silence than you realize" (Ramana Maharshi).

> "If we look very deeply, we will transcend birth and death" (Thich Nhat Hanh).

> "Go, go, go beyond, go totally beyond, be rooted in the ground of enlightenment" (The Mantra of the Heart Sutra, translated by H. H. the 14th Dalai Lama).

Conclusions

Science has become the dominant force in most parts of the world. It affects our lives and society in countless ways. To avoid confusion with harmful consequences, we need an **understanding of science**, at least its general principles, to know what science can do for us and what it can't. In other words, we have to understand **the power of science, its limitations, and what lies beyond them**. If we are not aware of the limitations, we cannot see what lies beyond them and then our lives become impoverished and unfulfilled.

Widespread misconceptions about science persist in the general public and even among many scientists. These misconceptions may have grave and even disastrous consequences. Not being aware of them may be a matter of life or death as, for example, during the COVID-19 pandemic (Chapter 9).

One fundamental misconception about science concerns its **methodology**. It is often said that science can be defined by its method, which is therefore referred to as the scientific method. However, as Feyerabend has demonstrated, there is no single method that characterizes all science and has been followed by all scientists (Chapters 1, 9). Furthermore, science cannot be clearly delimited from non-science and from what is beyond science, such as personal insight, subjectivity, uniqueness, just sensing, philosophy, the arts, and spirituality. Hence, there is some overlap between science and the beyond, which I have indicated on the cover of this book with the overlap of the two faces of Science and Beyond.

Another misconception: the widespread assumption or belief that science can prove or disprove (refute) its tenets such as theories and laws. But **proof and disproof are not attainable in science. Uncertainty prevails** because we cannot know whether future observations and experiments will confirm what seems to be proven today. Hence, **science remains open-ended** (Chapter 2). **Replication** of observations and experiments cannot be guaranteed because of the **multifactorality of the context and the uniqueness of objects and events** (Chapter 3). And often **objectivity** cannot be attained because of many kinds of bias such as expectancy, selection, and confirmation bias, which distort reality (Chapter 4). **Subjectivity** tends to be considered negative in our scientific age, where objectivity reigns supreme. But although subjectivity may be more or less erroneous, personally and culturally conditioned, it can also have an enormous richness and may be the door to the most profound wisdom, such as the wisdom of the Laozi (Lao Tzu), the Buddha, and many other sages of the East and West. Defining subjectivity just in terms of bias, as it is often done, overlooks the deepest potential of subjectivity, namely that it may lead to wisdom, happiness, and peace beyond science.

Another misconception: **empirical evidence (facts) determines whether a theory and worldview are accepted and retained**. But adherents of a theory and a worldview such as the mechanistic materialist worldview often manipulate – consciously or subconsciously – empirical evidence in such a way that it fits their theory and worldview. Thus, **contradictory facts are often ignored, declared erroneous, or explained away through *ad hoc* hypotheses. Such manipulations render a theory and worldview immune to contradictory evidence.** However, as more and more contradictory facts accumulate, it becomes increasingly difficult to ignore them or explain them away. Nonetheless, Max Planck and others said that theories and worldviews disappear only with the death of their adherents, not through the accumulation of contradictory facts (Chapter 7). **Theories and worldviews are not only based on empirical evidence but also on the power of powerful individuals, organizations, communities, and subcultures. Therefore, Foucault referred to "power-knowledge," the convergence of power and knowledge (Chapter 7).**

Besides its manipulation through power-knowledge, **empiricism appears limited because it is based on sensory perception that involves thought and language. In contrast, just sensing, a direct experience without the interference of the mind and language, may open the door to the infinite beyond science that relies on thought and language.** As Korzybski and others have pointed out, **thought and language remove us to some extent from reality.**

Another misconception: Science is mechanistic and materialist by its very nature. But although **mechanistic materialist science** has yielded a large body of scientific results, it discloses only one way of seeing and understanding the world. Insisting that it is the only way limits science unnecessarily. It impoverishes the evidence on which science can be based because it excludes too many facts. As Kripal pointed out, **materialist scientists "have to deny, erase, and take off the table so much of human experience to retain the illusion of the completeness of the materialist model"** (Chapter 8).

In contrast, **holistic science transcends mechanistic materialism in various ways**: it may include subtle energies for which there is empirical evidence; it may include the feminine side of science; it may accept radical empiricism that includes inner experience in addition to the outer experience of mainstream science, all of which leads to broad science that may even include aspects of spirituality. Furthermore, there are holistic scientists who claim that consciousness is primary or that both matter and consciousness arise from a deeper reality. Such views allow to explain many phenomena such as extrasensory perception, which cannot be understood in terms of the mechanistic materialist worldview (Chapter 8).

Another misconception: science is value-free. But **science and society interact and thus values of society may be projected into science.** In the past, the Church dominated society. Nowadays science, especially materialist science, exercises much power. Thus, **scientific knowledge has become power-knowledge**. To a considerable extent, powerful materialist mainstream scientists and their community decide which evidence is acceptable. Their power augments further through cooperation with governments, corporations, etc. **Materialist science, capitalism, consumerism, and militarism reinforce one another.** In health care, the powerful medical establishment conspires with the pharmaceutical industry and government to suppress evidence-based alternative and integrative medicine as much as possible. During the COVID-19 pandemic, powerful medical scientists, backed by governments and big pharma, decided that physical distancing, masking, and hygiene were *the* solution to the crisis until a vaccine was developed. Alternatives founded on much evidence were ignored, dismissed, and suppressed as misinformation, disinformation, pseudoscience, and conspiracy theory. Thus, evidence is controlled by power-knowledge, and the consequences have been disastrous (Chapter 9).

Another misconception: science can give us the Truth. But if Truth is defined as *that which is*, **Truth remains unattainable in science for several reasons**:

a. Since scientific explanations and predictions are limited, they cannot fully reveal the Truth, that which is (Chapter 1).
b. Since science cannot provide proof, uncertainty about reality, the Truth, remains (Chapter 2).
c. Since science does not deal with uniqueness and since uniqueness is part of that which is, science does not deal with all that is, the Truth (Chapter 3).
d. Since science aims at objectivity, it tends to exclude subjectivity, which is part of all that is, the Truth (Chapter 4).
e. Working as a scientist, one knows that scientific research requires simplifications, which remove us more or less from that which is, the Truth (Chapter 7).
f. According to the semantic view of scientific theories and laws, the question is no longer whether theories and laws are true; the question is whether they apply (Chapter 7).
g. Since science uses **language, including logic**, and since everything expressed through language has been abstracted (selected) from reality, from all that is, science again cannot reach Truth. It can reach, at best, aspects of Truth (Chapters 5, 6). Philosophy, the arts, religion, and spirituality may reach other aspects of Truth that complement science and may go beyond science (Chapters 9-11). But they may also misrepresent Truth more or less, especially when philosophy and religion become caught in dogmatism and intolerance. Hence, we need religious *experience*, which may be considered **spiritual experience**, which may transcend

doctrines of the mind and reach deeply into the heart and the gut. Ultimately, we need an integration of head, heart, and gut.

If we don't see that science can give us at best only aspects of the Truth, we may get trapped in scientism, the belief that science delivers the Truth, which leads to a delusion that entails a degree of **insanity**. The medical and psychiatric profession usually limits the label "insanity" to cases of severe mental illness. However, a degree of insanity appears widespread in society, and the fact that it usually is not recognized as such makes it even worse. It continues to have many grave and even disastrous consequences such as conflicts, violence, and war (Appendix 2 and 3).

But being aware of the limitations of science, especially the unnecessary limitations of materialist science, can provide broader perspectives and open the doors to what lies beyond these limitations, including **philosophy (as the love of wisdom), the arts, and spirituality**. Hence, such awareness can be liberating for individuals and society. It has been my aim in writing this book to contribute to this **liberation, which may result in deeper insight, happiness, and peace. If we cannot see beyond science, our lives remain impoverished. Inasmuch as the beyond is expressed through language and logic, we even have to go beyond the beyond as pointed out in the Heart Sutra: "Go, go, go beyond, go totally beyond, be rooted in the ground of enlightenment"** (translation by the Dalai Lama 2005).

Despite this emphasis on the beyond, it would be unfortunate if this book would be seen as an effort to debunk science. I have worked as a scientist my whole life and have great respect and passion for science and its potential when it is pursued in an open-minded way that will also lead beyond science. As the title for this book, I did not choose "Beyond Science," but "Science *and* Beyond," which means that I want to embrace both science and what lies beyond science.

I understand *Science and Beyond* in a nondualistic sense: what is beyond science does not exclude science; it transcends and includes science. Similarly, mystery transcends and includes science, or as Albert Einstein wrote, science arises out of the mystery, which means that the mysterious is the source of science. Science, philosophy, art, and spirituality all arise out of this mysterious, unnamable source and thus present different aspects of reality.

In short, we can go beyond science through deep personal insight (in addition to science), subjectivity (in addition to objectivity), direct experience including just sensing, awareness and celebration of uniqueness that may reveal the unnamable mystery beyond language and logic (reason), which may lead to freedom (from the limited self, the ego) that we may experience in silence, contemplation, meditation, or spontaneously anytime, anywhere. Besides science, we have to value philosophy (as the love of wisdom), the arts, and spirituality that may lead to greater sanity and better health. To me, sanity is rooted in wholeness and holiness, in the

unity of science and what is beyond, the nonduality of the manifest (that is investigated in science) and the unmanifest (the source of everything), the nonduality of the named and the unnamable that, in the *Daodejing* (*Tao Te Ching*), is referred to as **"the darkest of the dark, the door of all mysteries"** (Sabbadini's translation).

> The unnamable is the eternally real.
> "Naming is the origin of all particular things.
> Free from desire, you realize the mystery.
> Caught in desire, you see only the manifestations.
> Yet mystery and manifestations arise from the same source.
> This source is called darkness.
> Darkness within darkness.
> The gateway to all understanding."
> (*Tao Te Ching*, Stephen Mitchell's translation)

Appendix 1
From Plant Morphology to Infinite Issues

Nowadays, the sciences have become highly specialized, which leads to a narrow focus. One can, however, transcend this narrow focus. In this appendix, I chose plant morphology to illustrate how one can go beyond it. Plant morphology, a subdiscipline of plant biology that focuses on the form of plants, has been my area of specialization, in which I carried out research for nearly forty years. So, I am familiar with this science, and I have learned how to go beyond it.

Agnes Arber, a preeminent plant morphologist of the twentieth century, pointed out that plant morphology "may seem a narrow road, but rightly conceived, it should, like other biological paths, lead us to infinite issues" (Arber 1950, p. 1). In this autobiographical sketch, I want to recount how, besides personal experiences, the investigation of plant form, **plant morphology, led me to broader, deeper, and infinite issues, to healing logic (fuzzy logic, both/and logic, Buddhist logic, Jain logic), oneness, wholeness, holiness, health, balance, complementarity, dynamics (process philosophy), non-identity (Alfred Korzybski), integral philosophy (Ken Wilber), laughter, silence, mystery, etc.**

In school and university, I learned that plants (such as flowering plants) consist of three fundamental kinds of organs: root, stem (caulome), and leaf (phyllome). A flower consists of a short stem and modified leaves (phyllomes). This is the tenet of what has been called **classical plant morphology**. It is **based on Aristotelian either/or logic**. Thus, any plant organ of flowering plants that we encounter must be either a root or a stem (caulome) or a leaf (phyllome). The question I asked myself early on in my career as a plant morphologist was whether nature actually follows this kind of logic, and I found that, in many cases, it does, and in others, it doesn't. Hence, Aristotelian either/or logic has limitations. It needs to be supplemented by **fuzzy logic, which leads to continuum morphology, and both/and logic, which implies the complementarity of contrasting or even contradictory views** (Rutishauser and Sattler 1985, Sattler and Jeune 1992, Sattler 1986, 1994, 2018, Rutishauser and Isler 2001, Rutishauser 2020). Besides plant morphology, these kinds of logic also contribute to a better understanding and greater sanity in many types of relationships and society. Their application could heal many conflicts and even prevent violence and wars. I therefore refer to them as healthy or **healing logic and healthy or healing ways of thinking. Buddhist logic** and **Jain logic** appear healthy and healing in an even deeper sense since they include either/or logic and both/and logic, and then lead us beyond logic to the **indescribable, mysterious ground of existence**.

According to classical plant morphology, the three fundamental kinds of organs are distinct as categories and within an individual plant. But continuum morphology has shown that there are intermediates between the categories that link the categories (Sattler 1986, 1994, 1996, Sattler and Jeune 1992, Cusset 1994, Rutishauser and Isler 2001, Rutishauser 2020). Furthermore, there is a continuum of the organs within an individual plant. A close inspection of a plant (such as a flowering plant) shows that we cannot find a line that separates the root from the stem and the stem from a leaf. Root, stem, and leaves form a continuum (as the colours of the rainbow form a continuum). Upon even closer inspection, we also fail to find a line that separates the plant from its environment: the root appears continuous with the soil and the stem and leaves with the air. Hence, the notion of the **soil-plant-air-continuum (SPAC)**. Furthermore, through the air, plants are connected with animals, including humans. Through solar radiation, plants are connected with the sun, and through cosmic radiation, plants are connected with the cosmos. Thus, everything appears interconnected in one all-inclusive whole. As the Nobel Laureate Barbara McClintock put it: "Basically, everything is one." Interestingly, quantum physics arrived at the same conclusion of **oneness** and **wholeness**. David Bohm referred to "undivided wholeness." This conclusion contradicts the way we normally experience the world as consisting of separate objects. We have been conditioned to see objects such as different plants, animals, or humans as separate from one another, which creates the basis for competition, conflict, and violence. If, however, we recognize the wholeness and oneness, the situation changes fundamentally because we realize that, by harming another, we harm ourselves, since the other is not separate from us but an integral part of the all-inclusive whole.

Wholeness is related to holiness and health. The three words have the same etymological root. But the relation appears much deeper than just etymology. An awareness of wholeness can create a feeling of awe that can be an expression of holiness. It can create a feeling of **sacredness** (see Ravindra 2000). For example, contemplating a flower can call forth a feeling of sacredness because of the interconnectedness of everything. William Blake expressed it poetically: "To see the world in a grain of sand and a heaven in a wild flower; hold infinity in the palm of your hand, and eternity in an hour."

Health can be seen as being in tune with the whole instead of being caught in fragments. Furthermore, in Chinese medicine, health is considered being in **balance**, which manifests itself physically, emotionally, and mentally (see Appendix 2). Working in academia convinced me of the importance of balance. Most academics seem to be caught in their favourite theory, paradigm, or worldview. Thus, most classical plant morphologists seem to be enslaved in the belief in the trinity of root, stem, and leaf, ignoring or rejecting other ways of conceiving of a plant such as, for example, Agnes Arber's partial-shoot theory of the leaf, which is based on fuzzy logic instead of the Aristotelian either/or logic of classical plant morphology (for other alternatives to classical plant morphology, see Rutishauser and Sattler 1985, Cusset 1982, Rutishauser 2020).

As I became more aware of alternatives in plant morphology, I learned that, in general, recognizing and acknowledging alternatives provides a more comprehensive view of reality. Even accepting apparently contradictory tenets can lead to a richer understanding when these tenets present different aspects of reality that complement one another. Hence, the importance of **the principle of complementarity that implies both/and logic**. This principle is well known from physics as the complementarity of the wave and particle view of electrons and light. It can also be understood in a very general sense and thus can be applied in many ways in science, art, politics, everyday life, etc.

In plant morphology, we can recognize the complementarity of models that subdivide the plant in different ways (Rutishauser and Sattler 1985, Sattler 2019, Rutishauser 2020), and we can endorse a complementarity between structural morphology (morphology in the traditional sense) and process morphology. According to structural morphology, plants consist of structures such as leaves, and processes occur within these structures, which implies a structure/process dualism that has been pointed out by Woodger already in 1929 in the first edition of his *Biological Principles* (see Woodger 1967, p. 330). **Process morphology** transcends this dualism because, in process morphology, structure itself is also seen as process: it is a very slow process that can be illustrated through time-lapse photography. In this view, a structure such as a leaf is a process combination: a combination of slow morphogenetic processes with faster physiological processes (Sattler 1992, 1994, 2018, 2019, Sattler and Rutishauser 1990, Jeune and Sattler 1992). In process morphology the continuum of plant structures becomes a dynamic continuum. On this view, flowers can be seen as a continuum of process combinations (Sattler 2017). I developed this idea first for the androecium, the male part of flowers (Sattler 1988). Earlier, I had already emphasized how changes in the position of primordia (heterotopy) require a more dynamic interpretation of the gynoecium, the female part of flowers, that does not rely on drawing imaginary lines of demarcation for which empirical evidence is lacking (Sattler 1974).

The recognition of the dynamic continuum has far-reaching consequences for the origin and evolution of land plants (Sattler 1998). It allows an integration of two opposite views: Zimmermann's telome theory and Hagemann's opposite view. According to the telome theory, land plants originated from radially symmetrical axes, called telomes, whereas according to Hagemann's view, they arose from dorsiventrally flattened structures. Thus, *either* axes *or* flattened (leaf-like) structures are taken as the ancestral pattern of land plants. This opposition is overcome through the dynamic continuum that recognizes a continuum between radially symmetrical axes and dorsiventrally flattened structures, which according to process morphology constitute process combinations. The fossil record supports this view of a dynamic continuum at the origin of land plant evolution (for references see Sattler 1998).

Process morphology represents a special case of a general **process philosophy** according to which everything in manifest reality is dynamic (Nicholson and Dupré 2018). As Heraclitus said long ago: "Panta rhei" (everything flows). Keeping this in mind, it seems foolish to hang on to anything that changes, as Alan Watts made clear in his book *The Wisdom of Insecurity,* which had a profound influence on my life. The Buddha also emphasized impermanence long ago.

Process connects everything, even opposites. In her last book, *The Manifold and the One* (1957), Agnes Arber dealt with many opposites such as the manifold and the one, becoming and being. She emphasized not only their complementarity but also their coincidence in a chapter entitled "The Coincidence of Contraries." We can become aware of this coincidence in mystical union beyond the thinking mind and may have glimpses of it in many situations.

With her deep morphological insights and wisdom, **Agnes Arber** had a great influence on me (Arber 1950, 1954, 1957, Sattler 2001a). One very important author I failed to study during my professional life was **Alfred Korzybski**. But after my retirement, I finally read his magnum opus, *Science and Sanity*, and, to my great surprise, discovered that as a result of my morphological research and personal experiences, I had arrived independently and through other authors at many of his conclusions, such as the following:

1. **infinite-valued logic (fuzzy logic)**
2. **process thinking** (as in **process morphology** and **process philosophy**)
3. **dynamic relativism (complementarity)**
4. **general uncertainty** (of which Heisenberg's uncertainty principle in physics is a special case)
5. **consciousness of abstracting**
6. **undefined terms**
7. **organism-as-a-whole-in-the-environment** (the integration of organism and environment)
8. **the map is not the territory**

However, reading Korzybski's *Science and Sanity* clarified additional issues for me. For example, it became very clear to me that, as Korzybski emphasized: **"Whatever you might *say* something "*is*", it is not."** Thus, when I say: "It is a flower," I know now that it isn't. Why? Because a flower is defined by a number of properties, but the actual flower is infinitely more than the number of properties selected (abstracted) in the definition. Even a picture is far less than the actual object. Magritte was well aware of that. He painted an apple, and above it, he wrote: "Ceci n'est pas une pomme" (This is not an apple) (see Chapter 6). Why not? Because it is only an image of an apple, which is far less than the actual apple. Therefore, instead of saying, "This is an apple," we say, "This is a picture of an apple." And instead of saying, "This is a flower," I say, "I call this a flower." What it really is, we don't know. It remains mysterious, beyond language. Language cannot capture reality as it is. Reality as it is can only be honoured in silence. Hence, **the importance of silence**, which is beyond science that relies on language and mathematics, a form of language.

Reading **Ken Wilber** has also had a great impact on me, but I am rather critical of him (Sattler 2008, 2014). In high school and university, I was already looking for "the big picture," a comprehensive map of the whole universe, including human existence. Therefore, I was delighted when I discovered Ken Wilber's **AQAL Map** (see, for example, Wilber 2007) that I found very comprehensive and of great significance for me as a plant morphologist and a human being. This map comprises four dimensions of experience. In a somewhat simplified form, Wilber distinguishes only three major dimensions: **self (art), nature (science), and culture (morals)**. Within these dimensions, Wilber recognizes different levels that form a **hierarchy,** which Wilber prefers to call a **holarchy**. In this holarchy, holons are holarchically arranged. Holons are entities such as atoms, molecules, cells, organs, organisms, etc. Classical plant morphology recognizes the holons root, stem, and leaf. However, in continuum morphology, these holons dissolve in the continuum of the whole plant (actually, they appear non-existent as separate entities from the start). Similarly, organisms dissolve (appear non-existent as separate entities) in the continuum of the organism-environment whole. Hence, a holistic continuum view undermines Wilber's holarchy, which is based on holons. However, to me, this does not mean that Wilber's holarchy is useless. It only means that it represents one view based on the fragmentation of reality into holons, whereas the continuum view is based on continuity. As I see it, these two views complement one another. Although the continuum view appears closer to reality, the fragmentation implied in the holarchical view seems not totally unrealistic. It seems to represent at least an aspect of the differentiation of nature: "roots" are different from "stems," and "leaves" are different from "stems" and somewhat articulated from "stems." Thus, classical plant morphology and continuum morphology represent different aspects of plants (Sattler 1996). They complement one another. Problems arise if one view is taken as the only correct one. Thus, when Wilber claims that manifest reality "is" holarchical, he denies the continuum and other non-holarchical views. But saying that manifest reality can be seen as holarchical does not exclude that it can also be seen as non-holarchical.

Thinking in terms of a hierarchy or holarchy and entities or holons that compose them is very common in our culture and science, and often it is taken so much for granted that it is seen as the *only* way one can think about almost anything. But alternatives enrich our view and understanding of reality. They include not only the **continuum** and **undivided wholeness,** but also **Yin-Yang, network thinking, and a mandalic view**. With these additional perspectives, one can go beyond Wilber's AQAL Map (Sattler 2008, 2014). I should add that Wilber also recognizes these alternatives to holarchical thinking, but not with regard to the most basic structure of the Kosmos. According to him, the manifest Kosmos *is* holarchical.

Insistence on only one view to the exclusion of others often leads to exaggerated defensiveness and seriousness that we can witness in many debates from conflicts in personal relationships to scientific, philosophical, ideological, religious, and ethnic antagonisms within and between nations. We can overcome such conflicts and antagonisms not only through the recognition of complementarity but also through a better sense of humour and laughter.

Laughter, often called the best medicine, can be relaxing and liberating because when we laugh, we cannot think at the same time, and therefore we transcend the thinking mind that is often intensely preoccupied with the defense of its one-sided intolerant way of thinking (such as hierarchy or holarchy). Toward the end of my career as a university professor, I sometimes laughed for a few minutes with my students. They were, of course, very surprised that a university professor, who is supposed to be serious, would engage in such wild, non-intellectual behaviour. But most of them realized how relaxing and beneficial it was for them, especially during highly stressful times before exams. Since then, after my retirement, I have led many laughter sessions, including **laughter yoga**. Again, most participants found these sessions beneficial in many ways, but some of them could not let go of their seriousness, which is not surprising because, in our society, we have been conditioned not to laugh for too long. But enlightened beings often have a great sense of humour and enjoy laughing. The Dalai Lama said: "I am a professional laugher."

Paradoxically, laughter can lead to **silence** because it short-circuits and transcends the chatter of the thinking mind. In silence we may enter the **mystery** of existence that opens beyond the realm of the thinking mind. As Albert Einstein put it: "The most beautiful thing we can experience is the mysterious. It is the source of all true art and science."

Picture of a passionflower that is a short stem with modified leaves according to classical plant morphology and a continuum of process combinations according to continuum and process morphology. Beyond classical morphology, continuum and process morphology, beyond space and time, and beyond language and science is the unnamable mystery, which we may experience in silence. Photo courtesy of Dr. Gabriele Werle.

> "If we could but understand a single flower we would know who we are and what the world is" (Tennyson).
>
> "If we could see the miracle of a single flower clearly, our whole life would change" (Buddha).

Acknowledgements – I want to express my profound gratitude to all those who have helped me on the way from plant morphology to infinite issues: my parents, teachers, colleagues, collaborators, friends, and students. I am grateful to my thesis director, Professor Herman Merxmüller, who allowed me to initiate research in plant morphology that subsequently led me to continuum and process morphology. I am deeply grateful to Dr. Bernard Jeune and Professor Rolf Rutishauser, who, through their collaboration with me, made the most valuable contributions to the further elaboration of continuum and process morphology. Among other scientists, artists, philosophers, and sages from whom I have learned much along my path, there have been too many to enumerate, but I have to mention at least Agnes Arber, whose writings have profoundly influenced me.

Appendix 2
Health and Sanity of Body, Speech, and Mind

Throughout the book, I referred to health and health care. This appendix elaborates on this topic.

CONTENTS
- Introduction
- Paths of the Body: Cultivating our three bodies - A healthy lifestyle - Conventional, alternative, and integrative medicine - Our natural and social environment - Emotional intelligence - Health and happiness - Integrating light and darkness - Humour, relaxation, and acceptance.
- Paths of Speech, Sound, and Silence: Healthy language-behaviour - Healing sounds - Silence.
- Paths of the Mind: Healthy thinking skills - An integral vision - Shadow work – The witnessing mind – Beyond the witnessing mind.
- Individuals and Society: 20 imbalances in society.
- Education

INTRODUCTION

Health involves balance and wholeness (see, for example, Weil 2004). It has been defined as "a state of well-being, resulting from a dynamic balance that involves the physical and psychological aspects of the organism, as well as its interactions with its natural and social environment" (Capra 1983, p. 323). Since health also implies wholeness, "healing is fundamentally a return to wholeness" (Chopra 2009, p. 38). And since wholeness may evoke a feeling of the holy or sacred, healing and health are also related to holiness (the sacred). To me, holiness means being one with wholeness. Not surprisingly, the words health, wholeness, and holiness share the same etymological root.

Sanity is often equated with mental health or soundness, which is related to sound (see below the section on Speech, Sound, and Silence). To me, sanity means understanding and being in tune with reality; in other words, not being deluded. Most people may not be able to liberate themselves from all delusions. So we may envisage degrees of sanity and degrees of insanity. Wholeness appears central to both sanity and health, but the concept of sanity usually tends to emphasize the mental aspect of health.

According to the above definitions of sanity and health, it seems that, in our society, most people and society itself suffer to various degrees from insanity and poor health. So, in this appendix, I want to show how we can gain more sanity and better health. My focus will be on the general population. Cases of extreme mental and physical illness have to be treated by specialists, but my suggestions may be relevant even to such severe cases.

There are many paths to more sanity and better health, including paths of the body, paths of speech, sound, and silence, and paths of the mind. I shall deal with them sequentially, keeping in mind that they are interconnected and grounded in the nondual.

PATHS OF THE BODY

Cultivating our 3 Bodies - In our culture, when we refer to the body, we usually mean the physical body. There is, however, evidence for the existence of subtle and very subtle (causal) bodies that represent energies. In some spiritual traditions such as Hindu Vedanta, those additional bodies or energies have been recognized for a long time. Nowadays, there is also scientific evidence for subtle energies that represent the subtle body (see, for example, Tiller 1997, 2007; Gerber 2001, Church 2018).

According to Greene (2009), all bodies, which can be seen as energies, form a continuum of energy frequencies. Divisions within this continuum may appear more or less arbitrary. However, often the above three bodies are distinguished. Some traditions and authors divided the continuum into more than three bodies (see, for example, Steiner 2002, Brennan 1988). Greene (2009) distinguished the following four bodies and their corresponding energies: the physical, vital, emotional, and universal. The vital and emotional bodies comprise the subtle body in the 3-body classification. The universal body corresponds to the very subtle (causal) body. "All of creation lies in your universal body. Don't be blinded by the personal aspects of your body so that you ignore the universal" (Nisargadatta in Hendricks & Johncock 2005, p. 37). "Your body and the universe are a single field of energy, information, and consciousness" (Chopra 2009, p. 27). This "awareness heals…and healing is fundamentally a return to wholeness" (ibid, p. 38). "Both Mahayana and Vajrayana Buddhism define complete awakening (Buddhahood) as the Trikaya, the union of physical, subtle, and super subtle [universal] bodies" (Powers 2016).

A Healthy Lifestyle - For good health, good posture and corporeal exercise are highly recommended. Good posture seems important for all three bodies, especially the physical and subtle bodies. It can be achieved through various techniques, such as the Alexander Technique, which is also helpful for improving the way we move.

A vast array of techniques is available for corporeal exercise. Whereas exercises such as weightlifting or aerobics strengthen mainly the physical body, yoga, Taiji (tai chi), Qigong, and other techniques also train the subtle body (subtle energies) and might even touch the very subtle (causal or universal) body. As part of an Integral Life Practice, Wilber et al. (2008) developed a 3-Body Workout that exercises all three bodies: the physical, the subtle and the very subtle (causal or universal) body.

Besides posture and corporeal exercise, **a healthy lifestyle**, **according to Dr. Weil (2004, 2011), also includes an anti-inflammatory diet, some supplements, reducing information overload, proper breathing and breathing exercises, rest, relaxation, sufficient sound sleep, stress reduction, meditation, laughter, silence, practising forgiveness, compassion, etc.** One should add **touch** to this list because it "is truly fundamental to human communication, bonding, and health" (Keltner 2010). Referring to "One Health," Lueddeke (2016) emphasizes the importance of a very comprehensive approach to health care and well-being. Avoiding pollutants, such as, for example, many chemicals that are also prevalent in the food we eat, cannot be overemphasized.

Conventional, Alternative, and Integrative Medicine - For specific debilitating diseases, Dr. Weil recommends, first of all, holistic alternative medicines, and if that does not help, conventional medicine. But, in some cases, the first choice might be conventional medicine, although it may have more or less severe negative side effects. Integrative medicine for the body-mind combines the best of conventional and alternative medicines (see, for example, Weil 2004, 2009). Unfortunately, most medical doctors seem to know little or nothing about alternative medicines and often denounce or ridicule them. Denouncing what one doesn't know to me is not a scientific attitude. Furthermore, as, for example, Lanctôt (1995) has pointed out, the conservative medical establishment, the pharmaceutical industry, and governments conspire to suppress alternative medicines.

An integrative approach to health and healing appears to be the solution to many diseases, including cancer, for which conventional medicine has had only minimal success despite enormous funding and research (see, for example, Somers 2009, p. XIX; Raza 2019). We need to rethink how we look at these diseases (see, for example, Whitaker in Somers 2009). We need to recognize that these diseases require "a multifactorial solution: medical, nutritional, and lifestyle changes, as well as mental, emotional, social, and spiritual issues" (Murray et al. 2002, p. XIV). We also have to include *Energy Medicine* (Oschman 2015) that understands the body as energy or energies. It is practiced, for example, at the National Institute for Integrative Healthcare (niih.org).

The recognition of the plasticity of the brain and its intimate interconnection with the rest of the body also plays an important role in healthcare. It has great potential for alleviating various illnesses, including chronic pain, stroke, Parkinson's disease, attention deficit disorder, autism, multiple sclerosis, traumatic brain injury, learning disorders, sensory processing disorders, balance problems, and, to some extent, dementia (Doidge 2015).

One also has to recognize that in addition to the brain in our head, there are remarkable neural networks in the belly. Some authors have referred to these networks in the belly as a "second brain." Even in the region of the heart, an accumulation of nerve cells has been referred to as a "heart brain." Therefore, a more comprehensive understanding of human neurology and health will have to include all three "brains," those in the head, the heart, and the gut.

In addition to the recognition of three "brains," we have to be aware of the living matrix that envelops and connects all organs of the body, including the nervous system. It consists of connective tissue that includes an extracellular environment, often referred to as fascia, which form a continuous network throughout the body, surrounding and permeating all tissues and organs (Oschman 2015, Chapters 10-12). The living matrix is considered a liquid crystalline semiconductor network in which communication is faster than in neurons (ibid., pp. 195-196). "Another potentially extremely rapid system that is just beginning to be explored is quantum coherence of water molecules" (ibid., p. 196). As a result, the organism can be seen as a coherent whole.

All of these findings lend support to the idea that the mind is not only located in the brain of the head, but the whole body and, as Sheldrake (2012, Chapter 8) and others have pointed out, also beyond. This more encompassing view of the mind has profound implications for health and healing. For example, it validates the relevance of meditation in which we transcend the limitations of individual consciousness and become aware of all-pervading fundamental consciousness (Blackstone 2008).

Insanity and mental illness, ranging from neuroses to more severe mental health problems, can be addressed through the mind and the body (or bodies). The use of drugs may be the last resort. Otherwise, there are many helpful methods, especially for less severe problems. For example, tapping certain points of the body may relieve various physical, emotional, and mental afflictions (Ortner 2013). Since the body is integrated with speech and the mind, it also seems of utmost importance to look into and transcend deeply ingrained harmful patterns of speech and the mind that can lead to much insanity and ill health. Meditation has been used successfully in this respect, but in certain cases it may not help or may make the condition even worse (Epstein 2007, p. 149). Otherwise, through meditation, we can become aware of the illusion that the ego (or self) represents an ultimate reality, and we can learn to overcome this illusion, which leads to more sanity - sanity in the sense of understanding and being in tune with reality.

Our Natural and Social Environment also matters for our physical, emotional, and mental health. Soil, water, and air pollution that affects the food we eat, the water we drink, and the air we breathe can lead to poor health. Therefore, the reduction of pollution is not only desirable for the environment but also for our health (see, for example, Epstein and Ferber 2011). Governments could reduce the escalating costs of health care if they took better care of the environment. Another way to reduce health-care costs is through the recognition of alternative medicines that are less expensive than conventional medicine and that, for many ailments (but not all) are often more effective than conventional treatments that often have negative side-effects.

The social environment of our society may also affect our health. Information overload, pressure, coercion, exploitation, stress, and other factors may contribute to poor health and insanity. These factors may affect our physical, emotional, and mental health.

Emotional Intelligence - Emotions are deeply rooted in the body and the mind, that is, thoughts. To deal with emotions, especially negative emotions such as anger, fear, or jealousy, it can be helpful to feel them in the body (see, for example, Cushnir 2008, Chopra 2009) and to investigate and observe mindfully the thoughts involved in them. Weil (2011) recommended various other aids, especially for people who suffer from depression, a common ailment in our society. In general, developing emotional intelligence can create more sanity and better health (see, for example, Goleman 2005).

Health and Happiness often appear interrelated. Healthy people often seem happier than sick people, and happy people with a positive outlook on life tend to be healthier than unhappy pessimists. A healthy lifestyle supports happiness. In addition, in his book *The Ultimate Happiness Prescription,* Chopra (2009) highlights seven keys for more profound happiness: 1. Body awareness, 2. Self-esteem that is rooted in the universal Self, 3. Shadow work, 4. Overcoming righteousness, 5. Focus on the present, 6. Seeing the world in oneself, 7. Living for enlightenment.

Individual happiness contributes to happiness in society and the world. And happiness heals. "The most important contribution I can make to the healing of our planet is therefore to be happy" (Chopra 2009, p. 138).

Integrating Light and Darkness – Happiness appears even more fundamental than health. Dossey (1984) suggested not to become obsessed with health because such obsession is not healthy. It appears healthier to accept unavoidable illness, to recognize that the light of health and the shadow of illness belong together like day and night, Yang and Yin. Through the acceptance of the opposites, one can go beyond them and possibly become aware of the very subtle universal body that manifests itself as very subtle energy or clear light in which we "recognize that space is light, that light is space, and that light and space are energy – there is no separation. This recognition of no separation appears as clear light. Clear light is not white, yellow, blue, red, or green. It is pure awareness. The moment you realize that light, you are liberated" (Wangyal Rinpoche 2011, p. 43).

Humour, Relaxation, and Acceptance - In our search of the ultimate, whatever that may be, it seems important to relax and allow, as a friend of mine would say, and appreciate the wisdom of humour and laughter that can be liberating because it transcends egocentricity (see also Appendix 1).

> "Randy Mustaver is telling his friend that he has toured around the whole world looking for a perfect woman.
> "Did you find her?" asks his friend.
> "Yes, I did," replies Randy. "But it is a sad story."
> "Why is that," asks the friend.
> "Well" says Randy, "she was looking for a perfect man." "(Osho 1998, p. 200).

PATHS OF SPEECH, SOUND, AND SILENCE

Healthy Language-Behaviour - Language and behaviour are intimately interrelated. Therefore, Korzybski (1958, 2010) referred to language-behaviour. He pointed out how commonly used language obscures and distorts reality, which may lead to unhealthy behaviour and various degrees of insanity (see Chapter 6). To overcome or at least reduce these harmful consequences of written and spoken language, he proposed several extensional devices. In their book *Drive Yourself Sane*, Kodish & Kodish (2011) discussed such devices as follows (see also Falconar 2000 and Stockdale 2009):

1. **Avoiding the "is" of identity and predication** will reinforce the awareness that whatever is expressed through language is not identical to what it refers to. The "is" of identity occurs in sentences such as "He is a liar," whereas the "is" of predication occurs in sentences such as "He is bad." Such sentences distort our perception of reality because what he actually "is" is much more than just a liar or just bad. To underline that he is much more, we would have to say, "He is a liar, etc." and "He is bad, etc." But if we don't want to add the "etc.," we would have to be at least aware of it and assume that the person(s) we talk to also have this awareness. But how many people can retain a constant awareness of the "etc."? In the absence of such awareness, insane and harmful behavioural reactions may occur. Korzybski referred to them as semantic reactions. To avoid or reduce the severity of such reactions, we need to **develop more awareness of what is left out in verbal statements, which leads to greater balance, sanity, and health.**

   Being unaware of the "etc." can also inflict much harm to oneself. For example, if I say, "I am a failure," and consider this as what I really am, my identity, I have a very distorted perception of myself, which may have harmful consequences. In reality, I am much more than just a failure. Keeping this in mind supports my sanity and health.

Instead of or in addition to the "etc.," we may say, for example, that "he appears dishonest to me," which underlines that this is just *my* view or perception and not a reality that exists independently of me.

2. **Indexing** - Although adding or at least being aware of the "etc." may diminish the insanity of the behavioural reaction, it is not sufficient. We also have to differentiate between different meanings of a noun such as "liar" or an adjective such as "bad." Korzybski referred to this differentiation as indexing, which means that we have to distinguish between $liar_1$, $liar_2$, etc., and bad_1, bad_2, etc. For example, $liar_1$ could be a liar who intentionally deceives people in order to profit from this deception, whereas $liar_2$ could be a person who lies to avoid unnecessary upset or harm. Similar distinctions can be made for the meaning of adjectives such as "bad." To draw attention to words with multiple meanings, Korzybski suggested placing them in quotation marks. If we are unaware of different meanings, **bypassing** may happen: we think we talk about the same thing when in fact, we don't, and such misunderstandings may have more or less unhealthy behavioural consequences. In addition, bypassing may happen when we forget that **meaning is not only in words but also in the people and their behaviour**, in other words, in non-verbal communication that plays a greater role than most people can imagine.

3. **Dating** - Even indexing is not sufficient to avoid unhealthy and harmful behavioural reactions. We also need dating, which means we have to add the date to statements such as "He is dishonest, etc." Maybe he was only "dishonest" last month or yesterday, but not today. Knowing this makes a big difference. People change in many ways. If language does not take into account the fluidity of reality, including people, it provides poor and rather misleading descriptions.

It has also been pointed out that nouns tend to imply statics, whereas verbs refer to dynamics. Since reality appears to be profoundly dynamic, a language consisting only or mainly of verbs seems closer to reality than our common languages that have a noun-verb structure (see Chapter 6).

4. **Context** - Adding the context also helps to render language more accurate. For example, instead of simply saying that he lied yesterday, adding that he found himself in a very stressful situation provides important, relevant information. The context seems important in all languages and especially in tribal languages. Chase (1938, p. 59) remarked that foreigners cannot learn a tribal language from books because it is a mixture of words and "context of situation."

5. **Interconnectedness** - Another reason why our common languages tend to provide a rather poor map of reality and thus may lead to unhealthy behavioural

reactions is because they tend to fragment reality into bits and pieces. They cut up wholeness, which I think is one reason we tend to lose the holiness, the sacredness of existence. For purposes of communication, it may be necessary to fragment when, for example, I want to tell my doctor that I feel pain in my ear. But we have to keep in mind that in reality my ear is not separate from the rest of my body, and my body is not separate from my environment. **To underline the interconnection and wholeness, Korzybski suggested connecting words by hyphens** such as the organism-in-its-environment, mind-speech-body, or language-behaviour. An unawareness of such connections has led to harmful consequences and seems at least partly responsible for the ecological crisis and much physical and mental illness.

6. **Avoiding absolutisms** - Another way to improve our language and reduce or avoid unhealthy behavioural reactions is to eliminate absolutisms and *allness* terms such as all, always, never, etc. Such absolutisms do not seem to correspond to reality, and, therefore, much injustice and suffering have been inflicted on people by their use.

7. **Qualifying and quantifying** may also help to reduce or avoid unhealthy and harmful behavioural reactions. For example, instead of an overgeneralization, we may say "as far as I know," or "to me," or "under these circumstances." Instead of making either/or statements, we may say "to some extent" or "to a degree," and we may use fuzzy log**ic** (see Chapter 5).

8. **Avoiding perfectionism** may also be helpful. Looking for perfect health, perfect love, perfect happiness, perfect peace, perfect truth, etc. seems unrealistic and futile.

9. **Pausing** - Sometimes it may be more appropriate to refrain from talking and instead change the ambiance through laughter, hugging, or whatever seems appropriate. But if we want to speak, pausing before we do so can be helpful. This way, we can consciously choose the most appropriate expression instead of more or less automatically following inappropriate habits. Furthermore, pausing may create a space in which we can directly relate to the unnamable ground of existence. Korzybski referred to the un-speakable.

Additional extensional devices have been proposed (Brooks and Brooks 2006). Implementing extensional devices could prevent much insanity, violence, and bloodshed. In some cases, it may even be a matter of life or death. I heard that, during a party, a woman referred to her partner as "He is weak." The man felt so insulted and became so enraged that he wanted to kill her after the party, and she barely managed to escape into a women's shelter. If she had followed the extensional devices, she would not have said, "He *is* weak," but might have said, "In some instances when he seems stressed because his boss treats him unfairly, he does not have the strength that he often displays." Or she might have formulated it somewhat differently without implying that he *is* weak, which means that his very being *is* weak. Weakness in some instances and certain contexts may have been one of his traits, but not his being, which is infinitely more than just one of his traits. What he really *is* cannot be told through language because language abstracts (selects) only some features from reality. Therefore, **"whatever you say a thing [or a person] is, it is not"** (Korzybski 2010, p. VIII) **because "whatever we might *say* belongs to the verbal level and not to the un-speakable, objective levels"** (Korzybski 1958, p. 409). Because of its abstract nature, language cannot reach the reality of existence that remains mysterious. And therefore, knowledge expressed through language and speech that convey meaning "cannot contain the immensity of life and existence… cannot contain the mystery of being" (Osho 2012: 05). Knowledge cannot reach that which is, reality or Truth (if Truth is defined as that which is).

To the extent that we have to use language, implementing extensional devices can create more sanity and better health. Other approaches can also be very helpful. Marshall Rosenberg's nonviolent communication (2003) addresses our relationships. Many (or most?) of our relationships seem to be more or less unhealthy (that is, out of balance) and more or less insane (that is, out of tune with reality). Rosenberg's non-violent communication can help to restore some measure of health and sanity. It involves empathy and communicating our needs instead of judging, criticizing, and attacking. Stone et al. (2010) also provided great help on how to improve communication. Gray (2012), Tannen (2001, 2011, 2015), and other authors discussed insights that may help to infuse more sanity into relationships between men and women.

Since language and speech usually imply some kind of logic, they relate to the thinking mind (see below), and since they often stir up emotions, they relate to the body-mind because emotions are rooted in both the body and the mind. Thus, the sanity of language and speech correlates with the mind's sanity and the body's health.

Healing Sounds - **Experiencing speech as sound (vibration) may bring us closer to the mystery of reality than clinging to the meaning of language**, which, although useful, remains limited (see, for example, Shinzen Young 2016). Any particular sound, such as the sound of a word or sentence, connects us to universal sound and thus may allow us to transcend the limitations of egoic experience and consciousness. Such transcendence may lead to more sanity (soundness) and better health because it seems more directly in tune with reality. Singing and chanting, such as chanting mantras, can also transcend the meaning of language, and therefore it has been emphasized in many spiritual and religious traditions to lead us toward the ground or source of existence. A mantra "is meaningless, it is just a pure sound" (Osho 1974, p. 359). Chant a mantra or any sound, and "this sound will become the door to all sounds. Hindus call it **Nada**

Brahma, the sound of all" (Berendt 1983). The benefits of music and music therapy have been known for a long time.

Silence - Sound is also powerful and empowering because it may lead us into silence. This may happen in different ways. For example, as we exhale with a sound, such as the sound of Om, we may enter into silence. Or, according to a Shiva meditation technique, silence can be found in the centre of sound, as we can find stillness in the centre of a cyclone (Osho 2010, p. 357). In silence, we can be all united and thus beyond the conflict of power-hungry egos. According to Schopenhauer, "the end of philosophy is silence" (quoted by Gardiner 1967:331). And Wittgenstein, whom many consider one of the most important philosophers of the twentieth century, concluded his *Tractatus Logico-Philosophicus* with reference to the unnamable, reminding us that "whereof one cannot speak, thereof one must be silent."

Silence appears basic. Out of silence arises sound. With sound, words are created and then thoughts as logical sequences of words. And with thoughts and observations are created science, philosophy, and religion, which constitute important facets of our civilization and culture. However, in our society, we tend to see science, philosophy, and religion as primary, and thus we tend to forget the basis of sound (vibration) and silence. This seems like putting the cart in front of the horse, which can lead to degrees of insanity and sickness. To regain more sanity and better health, we have to return to our roots: sound (vibration) and silence.

"Silence heals" (Ramana Maharshi in Hendricks and Johncock 2005, pp. 22-23). And "silence is the best language of all. More is communicated through silence than you realize... silence has a greater impact than preaching" (ibid., p.23). This has to be kept in mind when one writes or reads a book - ha, ha...

PATHS OF THE MIND

"Mind" can have very different meanings. Usually it refers to the thinking mind, but it can also refer to the witnessing mind and even the universal or cosmic mind.

Healthy Thinking Skills - Since Western society (that has infected many societies all over the world) places enormous emphasis on the thinking mind, let us see how we can gain more sanity and health through the thinking mind. The thinking mind uses concepts and logic that strongly influence our ways of thinking, which in turn influence and are influenced by our speech and social interactions.

Our predominant ways of thinking are still profoundly shaped by Aristotelian logic that implies the so-called laws of thought: the laws of identity, contradiction, and the excluded middle. As I have shown in Chapter 5, these laws are not in tune with reality, or only to a limited extent. If they are used where they don't apply, unhealthy and more or less insane behaviour results.

To avoid the inappropriate and harmful use of Aristotelian either/or logic, we need both/and logic, the principle of complementarity, and fuzzy logic. Buddhist and Jain logic can lead us even beyond logic (Chapter 5). We could live in a saner and healthier world if we paid more attention to these kinds of logic.

An Integral Vision - We could also alleviate much insanity and sickness if our concepts, conceptual frameworks, world views, philosophies, ideologies, and religions would be less one-sided, less dogmatic, less antagonistic, and more open. Ken Wilber (2001, 2006b, 2017) tried to overcome unnecessary conflicts and one-sidedness through his AQAL Map of the Kosmos, including human existence (see Chapter 8). Incorporating the AQAL Map into the school and university curriculum could reduce the insanity created by one-sided and dogmatic worldviews such as the predominant materialistic scientism of mainstream science and religious fundamentalism. But the AQAL Map should be taught in a critical and open-ended way so that it would not become an AQAL dogma (Sattler 2008, 2014).

Shadow-Work - Fortunately, Ken Wilber recognizes limitations of the thinking mind and the importance of shadow-work, meditation, and bodywork (Wilber et al. 2008). Shadow-work, a form of psychotherapy, can create awareness of unconscious thoughts and emotions that have been buried and disowned and then tend to be projected onto others. If they are made conscious and integrated, more sanity and better health may be gained.

The Witnessing Mind - There are many ways of meditation, all of which help to transcend the limitations of the thinking mind. One form of meditation that is increasingly used is called mindfulness (see, for example, Shinzen Young 2016). In mindfulness, we do not identify with the thinking mind; we observe or witness the thinking mind. This process of observation creates a distance through which obsessive thinking can be transcended. It then becomes clear that we are more than just our thinking mind and our thoughts. Thoughts *arise* within us, but we *are not* our thoughts. Similarly, we are not our emotions and body sensations. They, too, arise within us. This insight can create peace, more sanity, and better health.

The practice of mindfulness, which involves the witnessing mind, appears helpful in many ways in many aspects of life. For example, it can reduce stress (see Kabat-Zinn 2013). Teaching mindfulness in prisons has been beneficial to many prisoners. Teaching mindfulness in schools and universities could create a saner and healthier society (see Kabat-Zinn 2011). Ultimately, we can practice mindfulness in everyday life so that we will be mindful in any moment.

Beyond the Witnessing Mind – In mindfulness a separation or duality of the witness and the witnessed may still be present. However, this duality may be overcome when one experiences the witnessed as arising within the witness and the witness disappearing in the witnessed (Bodian 2017). This leads to a nondual dimension of being, the cosmic mind, which is also called undeluded mind (in Tibetan Buddhism) and divine mind (in yoga). "You have to disappear into it, like a dewdrop falling into the ocean. Your mind is like the dewdrop, the divine mind is like the ocean… The divine mind is absolute freedom – it is liberation from all limitations" (Rajneesh 1984). We can create this awareness in everyday life so that we realize that "each moment is the universe" (Katagiri Roshi 2007).

> "The most common unexpected injury most people suffer nowadays is being struck by an idea."(Rovin 1989, p. 80).

INDIVIDUALS AND SOCIETY

What I have referred to so far has been mainly intended to further individual health and sanity. As individuals become healthier and saner, society is positively affected. On the other hand, as society becomes healthier and saner, it becomes easier for individuals to improve their health and sanity. Obviously, there is a feedback between the health and sanity of individuals and society (see, for example, Fromm 1955/1990).

So far, only some pockets of our society have become saner and healthier. It seems, however, that the sanity of these pockets is spreading. Many organizations have been formed that emphasize a healthy unifying vision (Capra and Luisi 2016). Even in the business community of corporations and at universities, some steps have been taken here and there to reduce the insanity of materialism, capitalism, consumerism, and militarism (see, for example, McLaughlin 2005).

20 Imbalances in Society - However, despite some progress, the following unhealthy and insane imbalances will have to be overcome to gain better health and greater sanity (see also Fromm 1955/1990; Mintzberg 2015):
1. The imbalance between the arts and science/technology. In our society, enormous amounts of money are spent on science and technology and very little on the arts.
2. The imbalance between science and spirituality. More emphasis on spirituality, including meditation, could lead to a saner and healthier society.
3. Too much use of Aristotelian either/or logic and not sufficient awareness of Buddhist logic, Jain logic, modern forms of logic such as fuzzy logic and complementarity, which leads to tolerance of different and even opposite views.
4. Too much emphasis on knowledge and not enough on wisdom.

5. Too much emphasis on rational intelligence (IQ) at the expense of other kinds of intelligence, such as emotional and social intelligence (see, for example, Goleman 2005, 2006).
6. The imbalance between conventional and alternative medicines. Enormous amounts of money are spent on conventional medicine and very little on alternative medicine, which could be helpful in many cases where conventional medicine fails.
7. Exaggerated emphasis on growth and insufficient (though increasing) emphasis on sustainability.
8. Too much emphasis on the economy and too little on happiness. In Bhutan, "gross national happiness" has precedence over "gross national product," which is considered an indicator of "progress" in our society.
9. Exaggerated emphasis on quantity at the expense of quality.
10. Too much involvement in wars and violent confrontations and insufficient emphasis on peaceful resolutions of conflicts. Marshall Rosenberg's (2003) nonviolent communication, if more widely taught, could contribute to greater sanity and peace.
11. Enormous expenditures for the military and not enough for social well-being and reform.
12. Too much emphasis on punishment and not enough on understanding, compassion, love, and forgiveness.
13. Much consumerism and not enough emphasis on the wealth of the here and now.
14. Speed and hurry and not enough rest, repose, contemplation, and meditation.
15. Much competition and not enough cooperation.
16. Too many fragmented views and not enough emphasis on unifying visions of wholeness.
17. Preoccupation with technology and not enough abiding in nature (the natural world). Overreliance on digital information.
18. Much involvement in the profane and not enough appreciation of holiness and the mysterious, which can be all-inclusive as in nondualism.
19. Dominance of the written and spoken word and not enough appreciation of silence.
20. Too much identification with limited ideas and not enough emphasis on transpersonal and cosmic unity consciousness. Limited identities that have been constructed individually and socially need to be transcended so that we can be "at home in the universe" (Kauffman 1995), where we are all united (Hendricks and Johncock 2005).

> "It is no measure of health to be well adjusted to a profoundly sick society" (Krishnamurti).

EDUCATION

Unfortunately, our public education system does not sufficiently address these imbalances. A better educational system is needed. So far, only a few schools, colleges, and universities are providing better education. Among them are Waldorf schools, the Schumacher College in England, and Naropa University in Boulder, Colorado. At most other schools, colleges and universities, occasionally exceptional teachers can go beyond the prescribed curriculum that does not sufficiently address the imbalances I pointed out above.

In a comprehensive way, Osho (2019) described five dimensions of education:

1. Information.
2. Sciences.
3. The Art of living.
4. Creative arts.
5. Meditation and the art of dying.

Appendix 3
The Human Condition and Its Transcendence

We want to know how we have been conditioned. How does this conditioning limit us? How can we surpass it? How can we reach and embrace our fullest human potential? How can we reach profound and lasting happiness and peace?

Since science has become the dominant force in most parts of the world, the human condition has been much affected by science. But it goes deeper than science. Thus, we have to go beyond science to understand the human condition. This appendix traces the origins of the human condition, its evolution, and shows how it can be transcended.

CONTENTS
- Our Animal Ancestry
- Human Acquisitions: Formation of large groups - Words and language - Ideas - Logic - Thought, emotion, and the body - Conflicts - The amplification of primitive traits
- Three Stages of Human Evolution: primeval, philosophical, and scientific thinking
- Power-Knowledge: The convergence of power and knowledge
- Modern Myths (Beliefs)
- Transcendence of the Human Condition
- Transcendence in a Nutshell
- Transcendence in One Phrase
- Happiness
- The Future of Humanity

Enlightened shamans, the Laozi (Lao Tzu), the Buddha, Heraclitus, and other sages of East and West appear to have transcended the human condition. Still, the **vast majority of humans seem to be more or less conditioned by our animal ancestry, language** and **thought (ideas and logic)**.

OUR ANIMAL ANCESTRY

To understand the human condition (how human existence has been conditioned), it is not sufficient to examine human history as Hannah Arendt (1959, 1998) has done in many ways; we also have to look at human prehistory, including our closest relatives in the animal kingdom, the common chimpanzee and the bonobo (also called pygmy chimpanzee). Since humans are very similar to these two species of chimpanzees, it has been concluded, "we are just a third species of chimpanzees" (Jared Diamond, quoted by Hands 2016, p. 532). **We resemble chimpanzees genetically, physiologically, and above all, in our behaviour. However, we have larger brains that engender the capacity for language, reasoning, and insight,** culminating in the "**cognitive revolution**" (Harari 2014, Part One). As a result, some people such as the Laozi, the Buddha, and other sages in the East and West have become enlightened, but others practise torture and other forms of extreme cruelty worse than the most violent behaviour of chimpanzees. **Our larger brains have also led to the development of science and technology,** which have garnered important insights and applications but have also produced most of the worst global problems such as overpopulation, climate change, pollution of earth, sea, and air, destruction of natural habitats, rapid species extinction, modern warfare, the threat of nuclear war, the meltdown of nuclear reactors, cybercrime, and extreme inequalities of wealth and power in the capitalist system. Thus, Harari concluded: "Unfortunately, the Sapiens regime on earth has so far produced little that we can be proud of" (Harari 2014, p. 415).

According to the primatologist de Waal (2005), chimpanzees practice reciprocal and genuine altruism, help and console each other, have a sense of group belonging, mediate and forgive each other, but they also tend to be egocentric, power-hungry, greedy, hierarchical, competitive, territorial, xenophobic, aggressive, engage at times in cruelty (even sadism), and practise at times deadly warfare and infanticide, whereas bonobos tend to be more playful, egalitarian, and do not engage in deadly warfare and infanticide; bonobos appear relatively peaceful, highly sexual and sensual. Although humans can exhibit typical bonobo traits, it seems that we have more in common with chimpanzees. Thus, we have the propensity for selfish and unselfish behaviour. De Waal concluded that **like chimpanzees, "we are born with a gamut of tendencies from the basest to the noblest"** (de Waal 2005, p. 237). **This gamut of tendencies becomes exaggerated positively or more often negatively through language, ideas and logic, science and technology**, as I shall point out below.

Being power-hungry, territorial, and xenophobic, when two communities of chimpanzees encounter another one, they make war and kill. Similarly, human tribes tend to engage in warfare and killing. And the same happens in modern nations in even more cruel and devastating ways. Furthermore, the tribal mentality can be seen in many kinds of groups: ethnic, ideological, religious, political, etc. (see, for example, Goldberg 2018). Chua (2018) referred to "political tribes" such as, for example, the Republicans and Democrats in the United States. Have we evolved? De Waal (2005, p.141) concluded that "humans share intergroup behaviour with both chimps and bonobos. When relations between human societies are bad, they are worse than between chimps, but when they are good, they are better than between bonobos. Our warfare exceeds the chimpanzee's "animal" violence in alarming ways [because we can use science and technology and may be driven by ideas]. But at the same time, the payoffs from neighbourly relations are richer than in bonobos."

Contrary to chimpanzees, we have not only exploited and killed other members of our species but to a great extent, we have also ruined our environment, which has led to an ecological crisis. The greedy, power-hungry hierarchical orientation of our inner ape has become magnified through the thought of superiority over nature. In the West, this thought has been reinforced by the religious doctrine expounded in Genesis 1 of the Bible, where we are told by God to subdue the earth and to have dominion over everything. White (1967) sees the roots of our ecological crisis in this hubris. Continuing on this path might eventually lead to the demise of humanity. But better education and self-inquiry may still bring out more of our highest potential: cooperation, compassion, understanding, wisdom, and reverence for the mystery of nature (see below).

Philosophers, theologians, scientists, and laypersons have discussed human nature almost endlessly. Among other questions, they have asked whether humans are good or evil. If we consider our animal ancestry, it becomes obvious how misguided such discussions may have been. As research by de Waal and others has shown, we have the propensities for the traits we find in chimpanzees and bonobos, which means that we have the propensities to be good and bad, loving and destructive, egocentric and altruistic, etc. (see above). Which of these propensities become actualized depends to a great extent on environmental influences. Therefore, education becomes of prime importance. The verb 'to educate' literally means 'to bring out.' Thus, a good education will bring out good traits, whereas a bad education will lead to bad results. Above all, good education can lead to the transcendence of the human condition (see below), which is beyond opposites, such as good or evil, perfect or imperfect, beautiful or ugly, but includes these opposites in a nondual sense.

HUMAN ACQUISITIONS

In this section, I will point out the relevance of the following: formation of large groups, words and language, ideas, logic, thought, emotion, the body, conflicts, and the amplification of primitive traits.

Formation of Large Groups - Whereas chimpanzees live in small troops (up to several dozen individuals), humans formed very large groups: **tribes, nations, and empires.** Therefore, if a chimpanzee troop attacks another, only relatively few individuals may be injured or killed. In contrast, if a human tribe or nation attacks another one, thousands or millions of individuals may be injured or killed. Furthermore, the magnitude of bloodshed in humans also increased through science and technology (see below). Language and ideas that form cultures played an important role in the formation of large groups in humans.

A sense of belonging to a group already present in chimpanzees in humans can extend to larger groups such as tribes and nations. Furthermore, it may also include organizations, religions, ideologies, and other affiliations. It creates in-groups that may oppose out-groups and may lead to conflict and war between the in-groups and out-groups. Through insight and meditation, one can go beyond the opposition of the in- and out-group (see below).

Language – **Words and language played a crucial role in human evolution**. Words and other symbols, such as sounds or drawings, can help communication and orientation in the world. For example, the word "Toronto" indicates a location in space. If I say, "I go slowly to rainy Toronto," I inform people around me: "I" (a pronoun) indicates that it is me and not John, "go" (a verb) indicates an action, "slowly" (an adverb) specifies the action, "Toronto" (a noun) indicates where I am going, and "rainy" (an adjective) specifies the noun. Thus, a noun (Toronto), a pronoun (I), a verb (go), an adverb (slowly), an adjective (rainy), and a preposition (to) convey information. Other word classes such as conjunctions (and, but, because, etc.), determiners (a, the, many, etc.), and exclamations add further information. Thus, grammar may become more or less complex. Furthermore, language may become more or less integrated with our mind, emotions, body, and our culture, including myth, art, religion, science, and technology that may provide the medium for the expression of spoken and written language (see McLuhan, who coined the slogans "the medium is the message" and "the medium is the massage"). Our mind, emotions, body, culture and technology may then become the context for the understanding and interpretation of words and sentences. But language, mind, emotions, body, culture and technology may be best seen as interdependent. And thus one may conclude that "language is a subject of infinite complexity" (Izutsu 2012, p. 1).

Although useful, words and language fragment the world. "**Language divides**" (Osho 2010, p. 768). Words are distinct, but their referents are not. For example, the word "tree" is distinct, but what it refers to is connected to its surroundings. Similarly, the words "I" and "you" are distinct, but what they refer to are not separate entities but integrated parts of the world. Like the word "I," words that refer to ideas such as capitalism or communism refer to abstractions from reality. If this is not understood, then words and language can give us a fragmented and misleading view of the world and ourselves. Therefore, we have to keep in mind: "**Reality is far from words** and it is very different from what a naïve person thinks it is" (Falconar 2000, p. 7). Referring to Korzybski, Falconar (ibid, p. 6) wrote: "**Whatever we say it is, IT IS NOT**" because "whatever we say, something is always left out" (Rajneesh 1978, p. 6). If we say light, darkness is left out; if we say life, death is left out; if we say order, chaos is left out, etc. "To name a thing is to separate it from the rest" (ibid, p. 31). "Words are hard, solid, they cannot contain the opposite. Existence is liquid. It has the quality of containing the opposite within itself" (ibid, p. 33).

We often see the universe as a universe of things, objects, when it is a "named" universe. But "things do not preexist consciousness: they emerge in the act of naming" (Sabbadini 2013, p. 35). Things and naming are interdependent.

"Words are the main obstacle on the spiritual path, which can be seen in pure intellectuals…Their whole life is of words, so they become alienated from reality" (Falconar 2000, p. 40). Therefore, Aldous Huxley (1977, p.171-172) concluded: "Language helps us and destroys us… it allows us to do in cold blood the good and the evil." "The enlightened person …lives in language and then goes beyond it" (ibid, p. 173), goes beyond name and thought. Who can see beyond language and thought can see that there is no separate self and that there are no separate objects. But we have been conditioned to believe in a separate self and separate objects. This conditioned belief appears to be at the root of the human condition and the ecological crisis. It creates a separation of ourselves from others, separation from nature, and leads to competition, control, domination, and their destructive consequences such as aggression, violence, war, and the destruction of our environment (see, for example, Hutchins 2014).

Wildwood (2018) describes nineteen levels of separation that include language, settlement and farming, possession and property, organized religion, patriarchy, bureaucracy, writing, Reformation, science and industrialization, colonialism and capitalism, mechanistic thinking and modern technology.

Ideas - The human situation becomes most troublesome when we use words and language that refer to ideas, and when we believe in and identify with these ideas. For example, the belief in and identification with ideas such as a separate self, capitalism, communism, and fascism have had devastating consequences. Since chimpanzees do not seem to have ideas, they are not plagued by such devastating consequences. Chimpanzees may fight and kill for a fruit tree; humans do not only fight and kill for a precious mine but also for ideas. Thus, **ideas in human society have enormously magnified the instinctual propensity for aggressive competition and war that we inherited from chimpanzees** (more correctly, the lineage that gave rise to chimpanzees and humans). For example, think of the ideas that led to the holocaust, Stalin's communism, the Cultural Revolution in China, etc.

Harari (2015) pointed out how stories based on ideas shaped human history when these stories were shared by a large number of people. As pointed out above, "the crucial factor in our conquest of the world was our ability to connect many humans to one another" (Harari 2015, p. 153). According to Harari, this ability has been one of the most important human acquisitions because only in this way could ideas have a major impact on humanity.

I do not want to imply that all ideas are harmful. We can think of noble ideas such as truth, goodness, and beauty. Social activists are often inspired by noble ideas. But often they are so much possessed by these ideas (that are expressed through words and language) that they ignore or forget the wider context. This may have catastrophic consequences. For example, deposing a brutal dictator may seem like a great idea, but if subsequently the society is not capable of collaborative and peaceful functioning, it may end up in the worst chaos, as we have seen in Libya and other countries.

Furthermore, **noble ideas may lead to opposite ideas, and this may again engender conflict, aggression, and war** (see below). There is, however, another way. Instead of seeing opposites as antagonistic, we can see them as complementing one another. Then one can embrace or at least tolerate both and thus conflict, violence, and war can be avoided. Unfortunately, this has not happened often in our culture because we have been indoctrinated and conditioned to think in terms of either/or, not in terms of both/and (see below).

In addition to complementarity and both/and thinking, one can achieve a synthesis of opposites (opposite ideas) that reveals an underlying unity. This unity transcends opposites as a thesis and its antithesis are transcended by their synthesis. Furthermore, the unity can be seen as the coincidence of opposites, which "is held to lie beyond the reach of discursive thought, and to be revealed to intuitive insight alone" (Arber 1967, p. 77). "Goethe [the German poet and scientist] saw the coincidence of contraries everywhere… A pregnant instance is his assertion that truth and error are from *one* source, and that the destruction of error may often involve the destruction of truth" (ibid, p. 76). In Zen reference can be found to "the identity of black and white, or evil and good" (ibid, p. 75). This may be meant to jolt us beyond thought and the thinking mind.

Furthermore, as in Yin and Yang, in the real world, there seems to be at least some black in the white and vice versa. There appears to be some good in evil and vice versa, and thus again, the apparent antagonism becomes bridged. In *Siddhartha*, Hermann Hesse wrote: "Never is a man or a deed wholly Samsara or wholly Nirvana; never is a man wholly a saint or a sinner" (Hesse 1951, p. 115).

The unity of good and evil may also lead us beyond good and evil, which implies going beyond the *ideas* of good and evil (see also Nietzsche 2002). In this sense, the unity of opposites must be sought at a deeper or higher level than the opposites.

Ideas and especially the identification with ideas are propagated through words and language, which bestows enormous importance and influence on words and language. Yet, "Existence is beyond the power of words…From wonder into wonder existence opens" (Lao Tzu, Witten Bynner's translation). Paradoxically, this insight is expressed through words, but these words do not refer to an abstract idea but to an insight rooted in the experience or beingness of oneness (nonduality). Words and language that express ideas cannot contain the mystery of life and existence.

> The Tao is ungraspable
> How can her mind be at one with it?
> Because she doesn't cling to ideas.
> (Chapter 21 of the *Tao Te Ching*, translated by Stephen Mitchell)

> Out beyond ideas of wrongdoing and right doing
> there is a field
> I will meet you there.
> (Rumi)

Logic - **Logic is embedded in our language structure**. Different kinds of logic can be used. **The most commonly used logic is still Aristotelian logic, which is a logic of identity and either/or.** This kind of logic that is often used subconsciously because it is deeply embedded in our culture can lead to profound distortions of our perception of reality and thus may have devastating consequences. **Identity exists only in abstractions, in language, including mathematics, but not in the real world** of constant flux as Heraclitus and other sages have clearly understood long ago: I cannot step in the same river twice because I and the river have changed; neither I nor the river has remained identical. Even saying, "I am I," appears deeply misleading because I am not only I, I am also the universe, one with the universe, as many sages and mystics have understood and as even holistic science has recognized (see, for example, Hollick 2006).

Many people are looking for or are asserting an identity. They feel a strong urge to identify with something such as ethnicity, philosophy, ideology, religion, nationality, etc. But such identifications remain relative or limited and caught in oppositions: my ethnicity, philosophy, ideology, religion, or nationality versus yours. Such oppositions, reinforced by either/or logic, may lead to conflict, violence, and war unless their relativity is recognized; unless we realize that we *are* not this or that. As Korzybski has so clearly demonstrated through his Structural Differential, whatever you *say* you *are*, you *are not*. We need to recognize that we *are* infinitely more than what we can *say we are*. We partake in the unnamable mystery, which unites us. With this realization, the human condition that often implies ideas and identification with these ideas can be transcended because the mystery is not conditional (see Chapter 6).

I think that if Korzybski's insights, including his Structural Differential, would be widely taught in schools and universities, we would live in a much better world, a world with less conflict, violence, and war, a world with more tolerance, compassion, and happiness because we could see the relativity of ideas that divide us and because we could appreciate the unnamable mystery in which we are united. We would no longer mistake a map of images, words, and ideas with the territory of reality.

"La condition humaine" (The human condition) by René Magritte (1935)

In this painting, Magritte shows that an image of reality is not reality. When we portray reality through an image or language, we may at best capture some aspects of reality but not reality as it is. Images and linguistic representations can be seen as maps. Mistaking maps for what they represent leads to misunderstandings that may have grave or catastrophic consequences. Therefore, Korzybski emphasized so much that **a map is not the territory. Nonetheless, so many people continue confusing the map with the territory. This confusion characterizes the human condition to a great extent.**

A map is an abstraction from reality. Identification with an abstraction, which appears so common, removes us from reality and thus can create innumerable problems and suffering. To a great extent, **the human condition can be seen as identification with abstractions**.

Besides identification, thinking in terms of either/or also appears deeply embedded in our culture since Aristotle devised his either/or logic. This logic has dominated the world and continues to do so. Most of the time, most people, including most scientists, experience themselves and the world through the lens of Aristotelian either/or logic. This logic, although it may be appropriate in certain cases, tends to be unrealistic and highly divisive (see Chapter 5). An investigation of the world shows that situations often are not either/or, not black or white, true or false, good or bad, etc., but more or less in between these extremes, which means that we find much grey, partial correctness, partial falsehood, and people being good and bad to various degrees. But according to either/or logic, it must always be either this or that. Very often this means that if one claims to be right, the other must be wrong. As Hoggan (2016) put it: "I'm Right and You're an Idiot." This logic also reinforces the "us versus them" thinking. We can often see this kind of thinking in tribes, ethnic, religious, ideological, and political affiliations, nations, etc. (see Chua 2018). Such thinking is not conducive to profound understanding and a peaceful society. But it is part of the human condition. It can be overcome through different kinds of logic, such as Buddhist and Jain logic, that do not have the defects of either/or logic, including "us versus them thinking" (see below).

Thought, Emotion, and the Body - Thought comprises language, ideas, and logic. In an interview with Oprah, Eckhart Tolle noted, "**the human condition is being lost in thought**," that is, as we identify with it, we get lost in it and remain unaware of the mysterious source out of which thought arises. Since emotions can be understood as a combination of body sensations with thought, thought also affects our emotions, and again we may get lost in them if we are not aware that we are infinitely more than just our thoughts and emotions. In our universal existence, we may be likened to the ocean and thoughts and emotions of our egoic self to waves on the surface of the all-encompassing ocean. So often we tend to forget that we are like the ocean (the universal Self) and not just like the waves (the individual egoic self). Identifying with the waves (the ego) and thinking that this is all we are create the misery of the human condition (see, for example, Foster 2012). Since waves vanish, identifying with them creates the fear of extinction, the fear of death. This fear may be at the root of much aggression, violence and war. People often attack or go to war because they fear that the other may attack them.

Even the body and body sensations may be influenced by thought. For example, the experience of physical pain may be related to thought, to how we think about it. Having negative thoughts about physical pain and resisting it makes it worse, while having positive and accepting thoughts about it may alleviate it to a considerable extent. Our attitude appears to be crucial. Christopher Day (2007) published a book entitled "(my) Dying is Fun," in which he describes how his positive attitude about his extreme physical handicaps allowed him to have fun. For example, he described how getting dressed would be very difficult. Sometimes when he had almost succeeded, his pants would fall again, and he had to invent a contraption to lift them again. Instead of getting frustrated, he had fun finding solutions to his many challenges.

In her journey through extreme physical pain, Vidyamala Burch distinguished four phases: denial, bargaining, acceptance, and flourishing that came out of the acceptance. Denial and bargaining are very much influenced by the thinking mind, but the deepest acceptance goes beyond the wishes of the ego and leads to the universal Self, which transcends the human condition (see also Brach 2003, Foster 2012).

We can, of course, also be in the body without the thinking mind. Then the body may become the door to spacious awareness that transcends the individual self (see below).

Conflicts - The human condition may entail much conflict: conflict between selfless behaviour and selfish behaviour such as aggressive competition; conflict between more or less contradictory ideas and between different ways of reasoning; conflict between thought and emotions, the thinking mind and the heart. These conflicts can be resolved through deep insight and spiritual transformation. Using the ocean analogy that I described above, we can see that conflicts can be compared with waves of the ocean; we are not just the waves (the conflicts); we are like the ocean (or like the sky). The waves (the conflicts) have no separate existence; hence identifying with them appears delusory and thus creates suffering. But **transcending the identification with impermanent, transitory waves (conflicts) allows us to shift into freedom**. Kelly (2015) and other sages devised exercises that may facilitate this shift, requiring a profound spiritual transformation from the egoic self to the universal Self (see also below).

The Amplification of Primitive Traits - As I pointed out already, traits and behaviour found in chimpanzees have become enormously amplified in humans. For example, as in chimpanzees, humans can play power games between individuals and small groups, but in addition, humans get involved in power struggles between large groups such as tribes, nations, and empires. And the outcome of these struggles may be devastating at a scale unknown in chimpanzees. Instead of killing just another chimpanzee or another small group of chimpanzees, humans have the capacity to exterminate thousands and even millions because they have been able to cooperate in large numbers. Harari (2015, p. 153) concluded that "the crucial factor in our conquest of the world was our ability to connect many humans to one another." Language, ideas, science and technology have played an important role in this connection.

Amplification occurs not only with regard to the desire for power, domination, and the ability to kill but also with regard to altruism and compassion toward fellow humans. Religions such as Christianity have preached love, but unfortunately, good intentions have often been corrupted and undermined by power struggles.

THREE STAGES OF HUMAN EVOLUTION

In the evolution of reflective consciousness, Hands (2016) distinguished **three overlapping phases: primeval thinking, philosophical thinking, and scientific thinking**. All three phases, especially the first one, involved more than just thinking.

Primeval Thinking "is characterized by creativity, invention, imagination, and beliefs" (Hands 2016, p. 584). It involved much more than just thinking. It includes the wisdom of shamans, stories, myths, religions, and superstitions, some of which continue up to the present time (see Kovacs 2019).

The arts also arose during this early period of humanity and, in later periods, reached their highest expressions in creations such as, for example, Beethoven's ninth symphony and many other creations in music, poetry, and the visual arts. However, the arts may also degenerate into ugliness. And religions may degenerate into dogmatism and fanaticism that may become destructive.

Dogmatic religions can restrict or undermine our human potential and our lives in many ways. For example, sexual behaviour has been regulated to a great extent. In most societies, marriage has been considered a universal institution, and although more recently sexual and romantic relationships have become extended (see, for example, Kingma 1999), marriage (or at least monogamy) is still widely considered natural or remains at least an ideal. There are, however, societies and practices that have been far less restrictive in this respect. For example, in the Mosuo tribe in China (close to the Tibetan border), women do not marry, they may have as many lovers as they wish (or have only one), and consequently they do not even have a word for "husband" or "father." In a communal household, a woman has a private room where her lover(s) can visit her at night. Thus, her sex life is strictly voluntary, unimpeded by myths, religious dogmas, and philosophical tenets.

Tantra that originated in ancient India comprises much more than just sex. It can be seen as an integration of body, mind, and spirit. With regard to sex, it emphasizes the sacredness of sex and thus sex has been used as a path to enlightenment.

Tantra and other spiritual practices demonstrate that in addition to cognitive advances and the propensities we inherited from our animal ancestors, we acquired the potential to go beyond cognition, beyond the thinking mind, which led to the deep insights of shamans and sages such as the Laozi, the Buddha, and other sages of East and West. These insights foster transformations that transcend the human condition (see below).

Philosophical Thinking "was characterized by a desire to seek explanations that did not invoke imagined spirits or anthropomorphic gods or God...[It] most probably emerged first on the Indian sub-continent while the other centres were China and the Greek colony of Ionia. Philosophers used insight, often resulting from disciplined meditation, and reasoning, based on prior assumptions or interpretations of evidence" (Hands 2016, p. 540-541). Among the first Western philosophers were Anaximander and Heraclitus, whose philosophies are reminiscent of insights of the Laozi and Buddha who did not succumb to the fallacy of identity.

Although "philosophy" literally means "love of wisdom," since Plato, it has been mainly a love of ideas communicated through language, and thus the nonduality of body, mind, and spirit has often eluded us. Living mainly in one's head to the detriment of the heart and the belly (hara) creates an unhealthy imbalance that may have many harmful and even catastrophic consequences. According to Chinese medicine, health means balance, and imbalance means sickness. Hence, degrees of imbalance in our culture indicate degrees of sickness (see Appendix 2).

Plato's influence on humanity has been enormous. According to Whitehead (1929), "the safest general characterization of the European philosophical tradition is that it consists of a series of footnotes to Plato." And I would add that, by now, Plato's influence has been worldwide. Unfortunately, the common interpretation of Platonism represents only a very one-sided view of Plato's profound philosophy (Potari 2018). According to the common interpretation, the material world, including our body, is unreal or only a shadow of ideas (eternal forms or essences) that constitute ultimate reality. This view created an unnatural dualism between the mind and the body, and, by giving so much prominence to ideas, the stage was set for endless struggles between contradictory ideas. Thus, much misery has been created; many wars have been fought in the name of an idea, an ideology, or a religion that has been dominated by an idea or ideas. However, one could emphasize that many of Plato's ideas appear noble such as the ideas of the good, beauty, and truth. But as pointed out in the second chapter of the Laozi (*Tao Te Ching*), when we postulate the idea of the good, the idea of the bad, evil, arises; when we talk about beauty, ugliness arises; and when we extoll truth, falsehood arises. As I pointed out above, ideas create their opposites, and if opposites are not understood as a unity that transcends them, they may engender conflict, violence and war, the human condition, which has at least one of its roots in the dominance of the thinking mind (ideas, thought) over the unity of body, mind, and spirit. I think that most people are unaware of how much they are stuck in ideas and how much they are dominated by them because in our culture, we have been deeply conditioned by a common interpretation of Platonism that enormously overemphasizes the importance of ideas and the thinking mind (see, for example, Lent 2017). As a result, our propensity for playful, egalitarian sexuality that we share to some extent with bonobos was counteracted and could not sufficiently manifest except perhaps in some pockets of our society and in some societies remote from the platonic influence such as in the South Pacific where people lived natural, uninhibited sexuality before they were indoctrinated by Christian missionaries.

Aristotle, Plato's chief disciple, considered the human being a rational animal (according to many interpreters). Like Plato, he has had an enormous influence on the history of humanity and the human condition. In the Middle Ages, Thomas Aquinas, who considered Aristotle *the* philosopher, elaborated a highly influential church doctrine in line with rationalist Aristotelian philosophy. The earliest universities that were founded by the church also followed this orientation, and up to the present time, universities retain this rationalist bias (see, for example, Pinker 2018). Mystical insight is excluded from universities, except in very rare institutions such as Naropa University in Boulder, Colorado, which was founded by Chögyam Trungpa, a Tibetan Buddhist. Education at Naropa University comprises three components: academic study, contemplation, and community service.

As a result of the enormous influence of Aristotelian rationalism, Christianity and other religions, especially in the West have tended to be more concerned with doctrine than religious experience that goes beyond ideas to embrace the unnamable mystery of existence. But Aristotle's influence goes far beyond religion. His either/or logic and hierarchical thinking still remain our predominant ways of thinking in most mainstream sciences and in society (see below). The Daoist wisdom of Yin-Yang thinking has not yet been widely recognized.

In his painting, *The School of Athens*, the Renaissance artist Raphael depicted many philosophers of antiquity, all dominated by Plato and Aristotle in the centre under the arch.

I want to emphasize that I do not want to blame Aristotle and Plato for everything that went wrong in human history. I consider Aristotle and Plato important philosophers who have contributed much more than what seems reflected in common interpretations of Platonism and Aristotelianism. For example, Plato was also critical of his Theory of Forms (ideas), and Aristotle, besides his either/or logic, also pointed out the "more or less" (for Aristotle's Non-Aristotelianism, see Kodish 2013). Plato emphasized the realization of oneness, and Aristotle recognized universal consciousness out of which everything arises (Potari 2018). Both Plato and Aristotle referred to the unspeakable beyond words (Arendt 1959, p. 20). According to Campbell (1990, p. 124), "Aristotle's rationality was rational in its reference to something transcendent of rationality, but it has become increasingly strictly rational." And considering the human being as a rational animal, an idea often attributed to Aristotle overemphasized reason to the detriment of the heart, the gut, and the unnamable mystery that transcends the human condition (see, for example, Lent 2017).

Potari (2018) showed that Hellenism shared so much with Eastern wisdom of sages like the Laozi and Buddha that the often-noted East-West divide collapses. For example, "Socrates' description reminds us of the Tao "that cannot be told"; "the nameless" which is "the beginning of heaven and earth" "(Potari 2018).

> "The unnamable is the eternally real.
> Naming is the origin of all particular things."
> (Chapter 1 of the *Tao Te Ching*, translated by Stephen
> Mitchell; see also Sabbadini 2013, pp. 33-45)

One can only speculate how different European and world history would have been if, instead of the one-sided simplification of Plato's philosophy and Aristotelian logic, Anaximander, Heraclitus, and the nondual core of Plato's philosophy as pointed out by Potari (2018) would have been the predominant influence. I think it would have been very different and for the better in many ways.

Scientific Thinking—During the Renaissance and the so-called period of the Enlightenment - more appropriately called the Age of Reason – the emphasis of reason and observation (including the observation of experiments) led to the foundation of modern science, especially mechanistic materialistic science, which still dominates modern society. To a great extent, materialist science and technology have become a sacred cow, and Aristotelian logic still has a profound grip on the majority of people, although more realistic and more inclusive alternative kinds of logic such as both/and logic and fuzzy logic are available. Digital technology is based on binary (either/or) logic, and this technology will more and more dehumanize life and lead to the development of increasingly intelligent robots that eventually may dominate us and – as Stephen Hawking and others have warned – may lead to the demise of humankind – all this as the result of the dominance of reason and Aristotelian logic at the expense of Buddhist and Jain logic, which appear more realistic and much more inclusive because in addition to the "either/or," they include "both/and" and "neither/nor," the indescribable, the unnamable, the mystery beyond reason that transcends the human condition.

As a reaction to the one-sidedness of the age of reason, we had romanticism that emphasized feeling and emotion, thus highlighting that we are more than just reason. But romanticism could not curtail the dominance of reason and mechanistic materialist science. Although twentieth-century holistic science such as quantum physics, holistic biology, and holistic medicine have shown the limitations of Aristotelian logic and mechanistic science, up to the present time, mainstream science and medicine remain predominantly materialistic and mechanistic, and mainstream thinking still follows to a great extent binary either/or logic. The consequences of this kind of thinking, which, to a great extent, is perpetuated in schools, universities, and society, are well known: exaggerated competition, intolerance, conflict, violence, and war between individuals, groups, organizations, and nations.

Besides a profound challenge of mainstream thinking through more holistic approaches, the twentieth century has also brought about a renewed appreciation of the wisdom of the Laozi, the Buddha, and other sages in the East and West. Thus, we could learn many lessons from the twentieth century. Some people have learned them, but mainstream society still seems to be dominated by the mechanistic materialist worldview and Aristotelian logic. Korzybski founded Non-Aristotelian General Semantics, which recognizes the unnamable mystery of reality that transcends the human condition dominated by identification with ideas expressed through language. Non-Aristotelian does not mean Anti-Aristotelian. Non-Aristotelian General Semantics includes Aristotelian thinking but goes far beyond it (for a summary, see Korzybski 2010, pp. 182-183 and Falconar 2000, pp. 6-7). Teaching this semantics and other great insights of the twentieth century in schools and universities seems crucial, but to what extent will it happen?

To avoid misunderstandings, **I want to emphasize that I am not against mechanist materialistic mainstream science if it is balanced by holistic science and the recognition of the unnamable mystery.** Unfortunately, so far, mechanistic materialist science and technology and the belief that we are higher than animals and nature (humanism) dominate the world. "We are consequently wreaking havoc on our fellow animals and on the surrounding ecosystem, seeking little more than our own comfort and amusement, yet never finding satisfaction" (Harari 2014, p. 416). And if we continue in this direction, we risk destroying our environment and ourselves.

POWER-KNOWLEDGE - THE CONVERGENCE OF POWER AND KNOWLEDGE

Knowledge, such as scientific knowledge, gives us power, including (limited) power over the natural world. However, it also works the other way around, as Foucault has demonstrated in detailed historical investigations: power is used to define and control knowledge, power is used to decide which evidence is acceptable, power is used to decide what is science, and what is pseudo-science. Thus, the exercise of power that we inherited from our animal ancestors may determine or influence knowledge. For example, materialist scientists and the materialist scientific community use their power to further materialist science and suppress alternative research that does not subscribe to the materialist dogma. Thus, research is not just a free acquisition of knowledge based on objective evidence, but at least to some extent, a power game (see, for example, Sheldrake 2012/20). Scientific research and its results depend, at least to some extent, on the power of the scientific community and powerful individuals. **Thus, knowledge such as scientific knowledge appears deeply rooted in the power-hunger we inherited from our animal ancestors. The cognitive realm that is so characteristic of the human species is not independent of the will of power, and thus we are not just a rational animal.**

MODERN MYTHS (BELIEFS)

Myths can also exert much power. They can be seen as stories that, to some extent, shape our world, the way we perceive ourselves and the world and how we act on these perceptions. In *The Power of Myth*, Joseph Campbell recognized myths as "clues to the spiritual potentialities of the human life" (Campbell 1988, p. 5). However, in a less profound sense, the term *myth* refers to a popular *belief that is deemed false or questionable* (see also Kripal 2019, p. 133). What are modern myths in this sense? One prevalent myth is **materialism**, which has different versions. The most extreme version implies that matter is all there is. Its spiritual counterpart is the myth that consciousness is all there is or spirit is all there is, and that matter is illusory or non-existent. This myth can, however, be surmounted if consciousness or spirit is conceived as including matter (see, for example, Hoffman 2008, Wilber 2000c, 2017). Furthermore, spirituality can be grounded in spiritual experience that does not necessitate any belief in spirit. Similarly, religion can be grounded in religious experience that does not rely on any dogma.

To a great extent, science, which can be a very useful tool, has also become a myth when it is considered infallible and when it is believed that it will solve all problems of humankind (see, for example, Harari 2016, p. 271). The name of this myth is **scientism**. One can, of course, practise science and use its results without subscribing to the myth of scientism.

Mathematics that plays an increasingly important role in science and technology can also turn into a myth when it is believed that reality is essentially mathematical. This myth has ancient roots. Pythagoras claimed: "All things are numbers." And Galileo reiterated this myth in a different way when he said that the universe "is written in the language of mathematics." Needless to say, one can use mathematics without subscribing to its mythology.

Another myth is that we need **continued growth,** such as economic growth, which seems the basis of unchecked capitalism and consumerism (see, for example, Monbiot 2017). However, in the long run, an economy based on unlimited growth is not sustainable. It also leads to the exploitation of our environment and the environmental crisis, threatening our survival.

Two of the most recent myths have been referred to as **techno-humanism** (whose goal is to improve humans through technology) and data religion or **dataism,** according to which everything is a matter of data processing or information (Harari 2017, Chapters 10 and 11). Coupled with manipulative technology, dataism may lead to an info apocalypse ("infocalypse") in which news and fake-news may become blurred and interchangeable. "That future, according to Ovadya, will arrive with a slew of slick, easy-to-use, and eventually seamless technological tools for manipulating perception and falsifying reality, for which terms have already been coined — "reality apathy," "automated laser phishing," and "human puppets"… as Ovadya observes, anyone could make it "appear as if anything has happened, regardless of whether or not it did" (Warzel 2018). Schick (2020) referred to "deep fakes."

There are many other myths and beliefs that have shaped or influenced human existence and society. Since the fifteenth century, the West has dominated the world. According to Ferguson (2012), this domination has been due to an emphasis on the following **six strategies: competition, science, the rule of law (such as property rights), modern medicine, consumerism, and work ethic.** During the twentieth century, these strategies were exported to the rest of the world. As I pointed out already, competition has ancient roots in our animal ancestry. The rule of law and work ethic have religious roots. Science and modern medicine have roots in the cognitive revolution. And consumerism is related to capitalism, which emphasizes competition.

The most fundamental myth that seems to be at the root of all or most of the other myths: **the myth of separation**, a story that most people take for granted (see, for example, Eisenstein 2019). The deeper connection is often overlooked, especially in our mechanistic materialist society, and the consequences can be devastating.

> "We are like islands in the sea, separate on the surface
> but connected in the deep" (William James).

TRANSCENDENCE OF THE HUMAN CONDITION

Like chimpanzees, humans have the propensity for egocentricity, greed, hierarchism, competition (for power and influence), xenophobia, aggression, violence, and war.

However, like chimpanzees, we also have the propensity for cooperation, altruism, empathy, and compassion. Like bonobos, we even have the propensity to be playful, egalitarian, and peaceful. However, these propensities seem much less realized than those of chimpanzees. As I pointed out, several philosophical ideas that in the West stem especially from Plato and Aristotle have greatly counteracted a more peaceful life. People have fought over ideas, including religious ideas, and have even gone to war for their ideas. People have fought for their perceived identities, and using – consciously or subconsciously – Aristotle's either/or logic, have endlessly and often acrimoniously fought about whether something is either this or that, true or false, good or bad, etc. Thus, to a great extent, the human condition has been and continues to be a struggle based on identification with ideas that manifest and intensify the propensities for competition, aggression, violence, and war that we inherited from chimpanzees (more correctly, from the ancestral line that gave rise to chimpanzees and humans). Eckhart Tolle said that the human condition is being lost in thought. Others have said that our emotions create the human condition. But our emotions are influenced by thought. And even our body sensations may be influenced by thought. Thus, thought creates a filter through which we experience ourselves and the world. As a result, we become entrapped in a rather limited space and lose our freedom (Blackstone 2008, pp. 13 and 84).

It has been said that the often-unconscious belief in the story or idea of separation is at the root of the human condition: separation of oneself from others and nature. Such separation is based on the identification with an idea that is reinforced through language. Identification with other ideas, some of which I pointed out above, also plays an important role. Thus, in general, identification with thought or being lost in thought when we are not aware of this identification, limits and constrains us. But we are infinitely more than just ideas in our heads (see below).

With this understanding, it appears obvious how we can transcend the human condition: **Through education, we have to create more awareness that thought, that is, ideas are not ultimate realities but only abstractions from reality. Then we can make use of these abstractions, but we will no longer be identified with them and thus enslaved by them. We will be liberated.** Gurdjieff said:

Identification is the only sin (quoted by Osho 2010, p. 203).

Instead of being identified with and lost in thought, we can be aware that thought (including emotions and body sensations if they are influenced by thought) arise out of what Thich Nhat Hanh called the deeper self that is connected to the whole universe and what has also been called the source, no-thingness or emptiness (in the Buddhist sense), no-mind (in Zen), or spaciousness (see, for example, Kelly 2015). Some people have already attained this awareness or at least have had glimpses of it, but for most of us, it remains a great challenge because we tend to be more or less contracted in the small self, the superficial self, the ego, the false self. "To break through this false view is to be liberated from every sort of fear, pain, and anxiety" (Thich Nhat Hanh 1999). And this changes our behaviour. As we realize that everything is interconnected, we develop compassion because harming someone else means harming ourselves. For this reason, wisdom (of the interconnectedness) and compassion go hand in hand.

Western society and most other societies that have become infected by the West appear highly cerebral, which means being stuck in the thinking mind, the ego. Institutions of higher learning have reinforced this stuckness since the founding of the first universities in the Middle Ages under the influence of the church that propagated Aristotle's rational bias. Of course, we have a mind that can be useful in many ways. But we also have a body, and the body has profound wisdom if it is not manipulated by the thinking mind (see, for example, Tolle 2004, Chapter 6). During the last century and even more in this century, technology has been distancing us even more from our bodies (Harari 2018, Chapter 5). Hence, **one challenge for the transcendence of the human condition is a rediscovery of the body and the realization that, by going deeply into the body, we can transcend the individual body as we experience our profound connection to the whole universe** (see, for example, Blackstone 2008). It seems that Shamans have gained transcendence primarily through the body.

Kelly (2015) and others have devised exercises that allow us to glimpse our universal connection and cultivate it. Through his Structural Differential, Korzybski showed that, ultimately, we need to transcend reason and logic because of their inherent limitations. Most intellectuals probably resist such transcendence because they derive their (limited) security from reason and logic. Using not only either/or logic but also Buddhist and Jain logic, logic itself could help them to transcend logic. And Korzybski's Structural Differential could help them to transcend the limitations of language and thought. Therefore, for intellectuals, Buddhist and Jain logic as well as Korzybski's Structural Differential may be the easiest door to the indescribable, the unspeakable, the mysterious that may be revealed in silence. **In silence, we may transcend thought and language, all opposition of ideas, emotions, and body sensations (when enmeshed with ideas) that have divided us for millennia and have led to immeasurable conflict, violence, and bloodshed.**

"**In peace and silence you grow**" (Sri Nisargadatta).

Having and cultivating good thoughts, although praiseworthy, is not yet the transcendence of the human condition because good thoughts, like bad thoughts, arise and vanish; they change in our world of impermanence. Furthermore, good thoughts are opposed to bad thoughts and thus conflict and violence that have plagued humankind for millennia continue. Therefore, to transcend the human condition, we have to go beyond all thought, good and bad, all emotions and body sensations as they are influenced and conditioned by thought. Laozi (Lao Tzu) understood that very well when he wrote in the second chapter of the *Tao Te Ching*: "When people see [think] some things as good, other things become bad... Therefore the Master acts without doing anything and teaches without saying anything" (translation by Stephen Mitchell 1988). Such action and teaching go beyond thought and, hence, beyond the human condition.

"Out beyond ideas of wrongdoing and rightdoing, there is a field. I'll meet you there.
When the soul lies down in that grass,
the world is too full to talk about.
Ideas, language, even the phrase each other doesn't make any sense." (Rumi)

One can distinguish different paths to transcendence, such as the path of the shaman, the path of insight or understanding (leading to knowledge and wisdom) and the path of love or compassion. In yoga, the latter two paths are known as Jnana Yoga and Bhakti Yoga (devotional yoga). Wisdom, such as the understanding of the interconnectedness of everything (referred to as emptiness in Buddhism), leads directly to love or compassion. If I am connected to you and to nature, I will not harm you because you and I are one; nature and I are one. Love and compassion also entail oneness. The problem is that love can become easily corrupted and contaminated by ideas. Christians have tortured and killed others who would not subscribe to their idea of God and love. Romantic lovers often tend to be "self-centred and self-indulgent. Even as the romantic lover worships his beloved, he only worships himself" (Feuerstein 2006, p. 51). Nonetheless, love and compassion have the potential to lead to a transcendence of the human condition. As I see it, love and compassion need to be anchored in wisdom to avoid the risk of corruption and contamination by ideas.

The arts may create awareness of the human condition and may also play a role in its transcendence. For example, listening to Beethoven's 9th Symphony and other great works of art may lead us beyond the human condition, at least temporarily.

Besides pleasure and joy, there is much suffering in this world. Understanding suffering appears crucial for the transcendence of the human condition. Why do we suffer? According to the Buddha, we suffer because of craving (neediness) and aversion (rejection). Life is tough. We often don't get what we want or what we think we need, and often we get what we don't want, and thus we suffer. However, if we can learn to accept things as they are, we can be at peace. Accepting things as they are is often seen as resignation. However, resisting what has already happened doesn't change what has happened. It just makes us suffer. Therefore, *Radical Acceptance* (Brach 2003) can help us to transcend suffering and the human condition. "Radical" means at the roots. Brach (2003) and many other meditation teachers, psychotherapists, and sages provide help and advice on how to achieve radical acceptance. "When we bring Radical Acceptance to the enormity of desire [and aversion], allowing it to be as it is, neither resisting it nor grasping after it, the light of our awareness dissolves the wanting self into the source" (Brach 2003, p. 154). Then desires, cravings, and aversions appear just like ripples on the enormous ocean of being (see also Foster 2012). Note that accepting what is does not mean that we should not work for a better future. We can influence the future, but we cannot change the past.

We share with chimpanzees joy and distress, pleasure and suffering, and other deep-rooted emotions. But because of thought, language, science, and technology, the magnitude of suffering in humans may far surpass that of chimpanzees. We can inflict enormous suffering on ourselves and others. But deep insight can help us to transcend suffering, at least to some extent or altogether. Can chimpanzees also have such insight?

According to Shinzen Young (2016, pp. 61-63), so-called primitive peoples in native cultures probably attained transcendence more easily than most modern people. "Whereas modern people struggle for years with the complexities of their wandering thoughts, native peoples could, by and large, become quite one-pointed after drumming or singing for hours. So the simplicity of daily life would tend to make it easy for people to enter samadhi. Indeed, we could say that the formal meditation techniques used by people in post-literate civilizations are just a systematic way of doing what our remote ancestors did relatively naturally every day" (Shinzen Young 2016, p. 62). Thus, a transcendence of the human condition may have been much more common among our remote ancestors than for us today. In the *Power of Myth* (1988), Joseph Campbell pointed out how native peoples, through myth and ritual led by shamans, could blissfully enter the mystery of existence. However, myth and ritual can become a hindrance when people do not see it as a means of transcendence but instead get more or less stuck in it (what Campbell (1990, 96) calls concretization). Fortunately, the mystery of existence can also be gleaned and realized directly without myth and ritual through aware breathing, laughter, penetrating insight, and many other ways (see, for example, Katie 2002, Brach 2003, Tolle 2004, Blackstone 2008, Osho 2010, Kelly 2015).

TRANSCENDENCE IN A NUTSHELL

We are not our thoughts, emotions, and body sensations (feelings). We are like the infinity of the sky in which thoughts, emotions, and feelings arise and vanish like transient clouds. If we are attached to and identified with these transient clouds (thoughts, emotions, and feelings), we become caught in the human condition, which means being conditioned by thoughts that may also influence emotions and feelings. **However, if we remain anchored in infinity in which thoughts, emotions, and feelings arise and vanish, we gain freedom** (see Kelly 2015: *Shift into Freedom*).

In other words, transcendence happens when, instead of exclusively identifying with something, we identify with no-thing, no-thingness, boundlessness and recognize things and events as abstractions from this boundlessness.

Become no-thing, and you will be all.

TRANSCENDENCE IN ONE PHRASE

"**Ending the story of separation**" (Charles Eisenstein).
"We are here to awaken from the illusion of separateness" (Thich Nhat Hanh).

HAPPINESS

Transcendence of the human condition leads to profound happiness, a state that is beyond ordinary happiness and unhappiness. In contrast, being in the human condition's bondage can provide at best temporary happiness that sooner or later vanishes and turns into unhappiness. We are temporarily happy if our desires are fulfilled, if we obtain what we want, and if we can avoid what we don't want. However, in a world of impermanence, we may sooner or later lose what we treasure and be confronted with what we try to avoid. For example, we may lose our good health, or we may lose our lover, and we may be threatened or attacked by a malicious person. Thus, if our happiness is based on external circumstances that are beyond our control, it may be more or less short-lived or impermanent as external circumstances are impermanent. This means that no object or person can give us lasting happiness. Lasting happiness can only come from within us. As we gain peace and acceptance, we may attain this profound happiness. It need not exclude the temporary happiness that is based on the fulfillment of desire and avoidance because if we accept the transitoriness of temporary happiness, it may lead to the profound happiness that transcends attachment (clinging) and aversion (see Chopra 2009).

"Our notions about happiness entrap us. We forget that they are just ideas. Our idea of happiness can prevent us from actually being happy" (Thich Nhat Hanh).

THE FUTURE OF HUMANITY

So far, transcendence of the human condition has been limited to few people. Unless the way of transcendence becomes widely taught and practised in schools and universities, this seems unlikely to change. However, if research in neuroscience produces technologies that facilitate transcendence, transcendence might be possible for a greater segment of society, provided that people use such technologies. The Dalai Lama and other spiritual teachers have expressed interest in the development of such technologies (see, for example, Goleman and Davidson 2017). In his book *The Science of Enlightenment*, Shinzen Young (2016, Chapter 11) called it "My Happiest Thought."

However, science and technology also progress in another direction. The development of artificial intelligence (AI) will produce computers and robots that will replace workers in many professions. This will lead to a "useless class" (Harari 2017, pp. 370-382) and an upper class whose members will still retain their profession and, using AI and biotechnology, may manipulate the "useless class." Thus, "just as Big Data algorithms [of AI] might extinguish liberty [of the "useless class"], they might simultaneously create the most unequal societies that ever existed (Harari 2018, p. 72). Furthermore, "what we should worry about most is the shift in authority from humans to algorithms, which might … open the way to the rise of digital dictatorships" (ibid., p. 43). And as computers and robots will become super-intelligent, eventually we may become enslaved by computers and robots (as we have domesticated and enslaved animals). Therefore, as I pointed out above, Hawking and other scientists have warned that "AI could spell the end of the human race." However, before this happens, we might become extinguished through nuclear war. Engaging in a global nuclear war would be sheer stupidity, but, as Harari (ibid., p. 182) pointed out: "Human stupidity is one of the most important forces in history." Albert Einstein said: Two things are infinite: the universe and human stupidity, and I am not sure about the universe.

In general, the survival of the human species appears threatened by several forces:

1. The strong power drive that we share with chimpanzees, which tends to lead to a Yin/Yang imbalance (too much Yang).
2. As a result of this power drive, rampant capitalism coupled with an obsession for growth that leads to the destruction of our environment.
3. Aristotelian logic based on identity and either/or thinking, which leads to a distorted view of reality, intolerance, conflict, and war.
4. Technology (including AI) that threatens the natural balance.

To overcome these threats, we will need a healthy balance of Yin and Yang instead of the Yang predominance that exaggerates the power drive. We will need a wiser use of technology and a more encompassing logic that also recognizes fuzziness (continuum), both/and (complementarity), and neither/nor, which leads to the unnamable mystery in which we can all be united so that we will end the story of separation.

References

Alexander, E. 2018. Consciousness and the shifting scientific paradigm. Paradigm Explorer 127: 3-8.
Alford, Dan "Moonhawk" et al. 2009. *The Language of Spirituality*. DVD.
All, L. 2012. *Off Balance*. Self-published.
Angell, I. O. and Demetis, D. S. 2011. *Science's First Mistake: Delusions in Pursuit of Theory*. London/New York: Bloomsbury Academic.
Arber, A. 1950. *The Natural Philosophy of Plant Form*. Cambridge: Cambridge University Press.
Arber, A. 1954/1964. *The Mind and the Eye. A Study of the Biologist's Standpoint*. Cambridge: Cambridge University Press.
Arber, A. 1957. *The Manifold and the One*. London: John Murray. Quest Book edition 1967 by the Theosophical Publishing House.
Ardagh, A. 2007. *Awakening into Oneness*. Boulder, CO: Sounds True.
Barke, J. I. 2020. Fear fatigue is more dangerous than COVID-19. https://www.americanthinker.com/blog/2020/08/fear_fatigue_is_more_dangerous_than_covid19.html accessed 20 October 2020.
Bateman, C. 2015. *The Mythology of Evolution*. UK: Zero Books.
Beauregard, M. 2012a. *Brain Wars*. New York: Harper Collins.
Beauregard, M. 2012b. Brain wars. Network Review, No. 110: 11-13.
Beauregard, M. et al. 2014. Manifesto for a Post-Materialist Science. https://opensciences.org/files/pdfs/Manifesto-for-a-Post-Materialist-Science.pdf
Berendt, J.-E. 1991. *The World is Sound. Nada Brahma*. Rochester, VM: Destiny Books.
Bhakdi, S. 2020a. CV-9Teen – hype & hysteria? Demystification of the nightmare! https://www.youtube.com/watch?v=h5FHQDpzkMw posted 28 March 2020.
Bhakdi, S. 2020b. Are we being told the truth about COVID-19? https://www.youtube.com/watch?v=ZnpnBYgGARE&feature=youtu.be, posted 11November 2020.
Blackstone, J. 2008. *The Enlightenment Process. A Guide to Embodied Spiritual Awakening*. St. Paul, MN: Paragon House.
Bodian, S. 2017. *Beyond Mindfulness: The Direct Approach to Lasting Peace, Happiness, and Love*. Oakland, CA: Non-Duality Press.
Bodian, S. 2020. The spiritual teachings of COVID-19. https://www.scienceandnonduality.com/article/the-spiritual-teachings-of-covid-19, accessed 15 November 2020.
Bohm, D. 1976. *Fragmentation and Wholeness*. Jerusalem: The Van Leer Jerusalem Foundation.
Bohm, D. 1981. *Wholeness and the Implicate Order*. London: Routledge & Kegan Paul.
Bohm, D. and Hiley, B. J. 1995. *The Undivided Universe: An Ontological Interpretation of Quantum Theory*. London: Routledge.
Bortoft, H. 1996. *The Wholeness of Nature. Goethe's Way toward a Science of Conscious Participation in Nature*. Hudson, NY: Lindisfarne Press.
Bos, A. 2017. *Thinking Outside the Brain Box*. Floris Books.

Bowden, J. 2020. Coronavirus patients in areas with high air pollution more likely to die: research. https://thehill.com/policy/energy-environment/491508-coronavirus-patients-in-areas-with-high-air-pollution-more-likely accessed 20 October 2020.

Brach, T. 2003. *Radical Acceptance. Embracing your Life with the Heart of a Buddha.* New York: Bantam.

Brennan, B. A. 1988. *Hands of Light. A Guide to Healing through the Human Energy Field.* New York: A Bantam Book.

Brooks, J. S. and Brooks, M. C. 2006. Some "new" extensional devices. Etc: A Review of General Semantics 65: 62-66.

Büller, H. R. et al. 2009. Double-blind studies are not always optimum for evaluation of a novel therapy: the case of new anticoagulants. https://onlinelibrary.wiley.com/doi/full/10.1111/j.1538-7836.2008.02848.x accessed 7 October 2020.

Burch, V. 2010. Living well with pain and illness. https://www.youtube.com/watch?v=3TUxs8HOCRo accessed 22 October 2020.

Bush, Z. 2020. The virome webinar replay. https://zachbushmd.com/virome-replay/, posted 10 September 2020.

Callison, C. and Young, M. L. 2019. *Reckoning: Journalism's Limits and Possibilities.* Oxford: Oxford University Press.

Campbell, J. with Moyers, B. 1988. *The Power of Myth.* New York: Doubleday.

Campbell, J. 1990. *Transformations of Myth through Time.* New York: Harper & Row.

Capra, F. 1975. *The Tao of Physics. An Exploration of the Parallels Between Modern Physics and Eastern Mysticism.* Berkeley, CA: Shambhala (5th edition 2010).

Capra, F. 1983. *The Turning Point. Science, Society, and the Rising Culture.* New York: Bantam (first published 1982).

Capra, F. 1996. *The Web of Life.* New York: Anchor Books.

Capra, F. and Luisi, P. L. 2016. *The Systems View of Life: A Unifying Vision.* Cambridge: Cambridge University Press.

Carter, C. 2012. *Science and Psychic Phenomena.* Rochester, VT: Inner Traditions.

Cathcarts, T. and Klein, D. 2010. *Heidegger and a Hippo walk through those Pearly Gates.* New York: Penguin Books.

Chase, S. 1938. *The Tyranny of Words.* New York: Harcourt, Brace, and Co.

Cheng, R. Z. et al. 2020. Early large dose of intravenous vitamin C is the treatment of choice for 2019-nCov pneumonia. Paradigm Explorer 132 (1): 6-8. (see also https://doi.org/10.1016/j.medidd.2020.100028).

Chopra, D. 2005. *Peace is the Way. Bringing War and Violence to an End.* New York: Harmony Books.

Chopra, D. 2009. *The Ultimate Happiness Prescription. 7 Keys to Joy and Enlightenment.* New York: Harmony Books.

Chopra, D. 2019. *Metahuman. Unleashing Your Infinite Potential.* New York: Harmony Books.

Chopra, D. and Kafatos, M. 2017. *You are the Universe. Discovering your Cosmic Self and why it matters.* New York: Harmony Books.

Chua, A. 2018. *Political Tribes.* New York: Penguin Press.

Church, D. 2009. *The Genie in you Genes: Epigenetic Medicine and the New Biology of Intention.* Nashville, TN: Cumberland House Publishing.

Church, D. 2018. *Mind to Matter*. New York: Hay House.
Cousins, N. 1981. *Anatomy of an Illness*. New York: Bantam Books.
Cox, C. 1999. *Nietzsche. Naturalism and Interpretation*. Berkeley, CA: University of California Press.
Csikszentmihalyi, M. 2008. *Flow. The Psychology of Optimal Experi*ence. New York: Harper Perennial Modern Classics (first published 1990).
Cushnir, R. 2008. *The One Thing Holding You Back. Unleashing the Power of Emotional Connection*. New York: HarperOne.
Cusset, G. 1982. The conceptual bases of plant morphology. In: Sattler, R. *Axioms and Principles of Plant Construction*. The Hague/Boston/London: Martinus Nijhoff/Dr. Junk Publishers, pp. 8-86 (also published in *Acta Biotheoretica* Vol. 31A).
Cusset, G. 1994. A simple classification of the complex parts of vascular plants. Botanical Journal of the Linnean Society 114: 229-242.
Dalai Lama [Tenzin Gyatso]. 2005a. *Essence of the Heart Sutra*. Boston: Wisdom Publications.
Dalai Lama [Tenzin Gyatso]. 2005b. *The Universe in a Single Atom: The Convergence of Science and Spirituality*. New York: Morgan Road Books.
Dalai Lama [Tenzin Gyatso]. 2011. *Beyond Religion. Ethics for a Whole World*. Boston: Mariner Books.
Dalai Lama [Tenzin Gyatso]. 2018. *Where Buddhism meets Neuroscience: Conversations with the Dalai Lama on the Spiritual and Scientific Views of Our Minds* (edited by Houshmand, Z. et al.). Boston: Shambhala.
Dalai Lama [Tenzin Gyatso] 2017. *Science and Philosophy in the Indian Buddhist Classics*, edited by Thupten Jinpa. Vol. 1: The Physical World. Boston, MA: Wisdom Publications.
Dalai Lama [Tenzin Gyatso] 2020. *Science and Philosophy in the Indian Buddhist Classics*, edited by Thupten Jinpa. Vol. 2: The Mind. Boston, MA: Wisdom Publications.
Dalai Lama [Tenzin Gyatso] and Cutler, H. 2009. *The Art of Happiness* (10th anniversary edition). New York: Riverhead Books.
Davis, I. 2020. A conspiracy theorist confesses. https://off-guardian.org/2020/06/03/a-conspiracy-theorist-confesses/ accessed 20 October 2020.
Dawes, M. 2010. Clearer thinking through practicing E-Prime. Etc: A Review of General Semantics 67 (4): 447-451.
Day, C. 2007. *(my) Dying is Fun. A Comedy of Disabled Adventures*. Victoria, BC: Trafford.
Dennett, D. C. 1995. *Darwin's Dangerous Idea: Evolution and the Meaning of Life*. New York: Simon & Schuster.
De Waal, F. 2005. *Our Inner Ape. A Leading Primatologist Explains Why We Are Who We Are*. New York: Riverhead Books.
De Waal, F. 2013. *The Bonobo and the Atheist*. New York: Norton.
Deutscher, G. 2005. *The Unfolding of Language*. New York: Metropolitan Books.
Diem-Lane, A. 2020. The many-sided brain. The Jain approach to studying consciousness. http://www.integralworld.net/diem-lane36.html accessed 22 October 2020.
Doidge, N. 2015. *The Brain's Way of Healing: Remarkable Discoveries and Recoveries from the Frontiers of Neuroplasticity*. New York: Viking.
Dossey, L. 1984. *Beyond Illness*. Boulder & London: New Science Library, Shambhala.
Dossey, L. 2000. *Reinventing Medicine. Beyond Mind-Body to a new Era of Healing*. New York: Harper One.

Dossey, L. 2013. *One Mind. How Our Individual Mind Is Part of a Greater Consciousness and Why It Matters.* Carlsbad, CA: Hay House.
Dupré, J. 2012. *Processes of Life: Essays in the Philosophy of Biology.* Oxford: Oxford University Press.
Edwards, J. 2007. *Choosing to Heal.* London: Watkins.
Edwards, P. (ed.) 1967. *The Encyclopedia of Philosophy.* Vol. 4, pp. 414-417. New York: Macmillan.
Einstein, A. 1932. My Credo. http://www.einstein-website.de/z_biography/credo.html accessed 22 October 2020.
Einstein, A. 1954. *Ideas and Opinions.* New York: Crown Publishers.
Eisenstein, C. 2013. A state of belief is a state of being. Network Review No.113: 3-6.
Eisenstein, C. 2019. Ending the story of separation. https://www.youtube.com/watch?v=Se5On6_PCSM accessed 22 October 2020.
Eisenstein, C. 2020a. The coronation. https://charleseisenstein.org/essays/the-coronation/ posted March 2020.
Eisenstein, C. 2020b. The conspiracy myth. https://charleseisenstein.org/essays/the-conspiracy-myth/ posted May 2020.
Epstein, M. 2007. *Psychotherapy without the Self. A Buddhist Perspective.* New Haven & London: Yale University Press.
Epstein, P.R. and D. Ferber. 2011. *Changing Planet, Changing Health. How the Climate Crisis Threatens Our Health and What We Can Do about it.* Berkeley & Los Angeles, CA: University of California Press.
Falconar, T. 2000. *Creative Intelligence and Self-Liberation. Korzybski, Non-Aristotelian Thinking and Eastern Realization.* Williston, VT: Crown House Publishing.
Ferguson, N. 2012. *Civilization: The West and the Rest.* New York: Penguin Books.
Ferrer, J. N. 2004. Integral transformative practice. http://www.integralworld.net/ferrer.html
Ferrer, J. N. 2017. *Participation and the Mystery.* Albany, NY: SUNY Pres
Feuerstein, G. 2006. *A Little Book for Lovers.* Boulder, CO: Sounds True.
Feyerabend, P. 1975a. *Against Method.* London: Verso.
Feyerabend, P. 1975b. How do defend society against science. https://www.radicalphilosophyarchive.com/issue-files/rp11_article1_defendsocietyagainstscience_feyerabend.pdf
Feyerabend, P. 1978. *Science in a Free Society.* London: New Left Books.
Feyerabend, P. 1987. *Farewell to Reason.* London: Verso/New Left Books.
Feyerabend, P. 2011. *The Tyranny of Science* (ed. by E Oberheim). Cambridge: Polity.
Firstenberg, A. 2020. *The Invisible Rainbow: A History of Electricity and Life.* Hartford, VT: Chelsea Green.
Forest, J. 2016. *The Root of War is Fear - Thomas Merton's Advice to Peacemakers.* Maryknoll, NY: Orbis Books.
Foster, J. 2012. *The Deepest Acceptance. Radical Awakening in Ordinary Life.* Boulder, CO: Sounds True.
Foucault, M. 1983. The Subject of Power. In: Dreyfus, H. and Rabinow, P. (eds.) *Beyond Structuralism and Hermeneutics.* Chicago, IL: University of Chicago Press, pp. 208-226.
Fromm, E. 1955/1990. *The Sane Society.* New York: Henry Holt & Co.
Gabriel, M. 2018. *I am Not a Brain. Philosophy of Mind for the 21st Century.* New York: Polity.
Gaille, L. 2020. 16 advantages and disadvantages of a double-blind study. https://vittana.org/16-advantages-and-disadvantages-of-a-double-blind-study, accessed 7 October 2020.

Gan, N. and Qiong,Y. 2020. Beijing is promoting traditional medicine as a 'Chinese solution' to coronavirus. Not everyone is aboard. https://www.cnn.com/2020/03/14/asia/coronavirus-traditional-chinese-medicine-intl-hnk/index.html, accessed 4 November 2020.

Gardiner, P. 1967. Schopenhauer, Arthur. In: Edwards, P. (ed.) *The Encyclopedia of Philosophy*. Vol. 7. New York: Macmillan Publishing.

Gelman, A. and Hennig, C. 2017. Beyond subjective and objective in statistics. Journal of the Royal Statistical Society, Series A, pp.1-67. http://www.stat.columbia.edu/~gelman/research/published/gelman_hennig_full_discussion.pdf accessed 8 8 October 2020.

Gerber, R. 2001. *Vibrational Medicine*. Rochester, VT: Bear & Co.

Gerretsen, I. 2020. How air pollution exacerbates Covid-19. https://www.bbc.com/future/article/20200427-how-air-pollution-exacerbates-covid-19, accessed 20 20 October 2020.

Giere, R. N. 1979. *Understanding Scientific Reasoning*. New York: Holt, Rinehart and Winston.

Giere, R. N. 2006. *Scientific Perspectivism*. Chicago: University of Chicago Press.

Glazier, J. W. 2019. *Arts of Subjectivity: A New Animism for the Post-Media Era*. New York: Bloomsbury Academic.

Glynn, I. 2013. *Elegance in Science: The Beauty of Simplicity*. New York: Oxford University Press.

Gober, M. 2018a. Shifting the paradigm of consciousness and why it is essential. Paradigm Explorer 128:19-21.

Gober, M. 2018b. *An End to Upside Down Thinking*. Sherfield on Loddon, UK: Waterside Press.

Goldacre, B. 2012. *Bad Pharma. How Drug Companies Mislead Doctors and Harm Patients*. Toronto: Signal.

Goldberg, J. 2018. *Suicide of the West. How the Rebirth of Tribalism, Populism, Nationalism, and Identity Politics is Destroying American Democracy*. New York: Crown Forum.

Goleman, D. 2005. *Emotional Intelligence. Why it can matter more than IQ*. New York: Bantam Dell.

Goleman, D. and Davidson, R. J. 2017. *Altered Traits: Science Reveals How Meditation Changes Your Mind, Brain, and Body*. New York: Avery.

Goodwin, B. 2001. How the Leopard Changed its Spots: The Evolution of Complexity. Princeton: Princeton University Press.

Goodwin, B. 2007. Nature's Due: Healing Our Fragmented Culture. Edinburgh: Floris Books.

Gorbachev, M. 2020. *What Is at Stake Now: My Appeal for Peace and Freedom*. Cambridge, UK: Polity.

Gould, S. J. 1973. *Ever Since Darwin. Reflections in Natural History*. New York: Norton.

Gray, J. 2012. *Men Are from Mars, Women Are from Venus*. 20th anniversary edition. New York: HarperCollins.

Greene, D. 2009. *Endless Energy. The Essential Guide to Energy Health*. Maui, HI: MetaComm Media.

Griffin, D. R. (ed.). 1988. *The Reenchantment of Science*. Albany, NY: State University of New York Press.

Grof, S. 2006. *When the Impossible Happens: Adventures in Non-Ordinary Realities*. Amazon.com Services LLC.

Grof, S. and Grof, C. (eds.) 1989. *Spiritual Emergency. When Personal Transformation becomes a Crisis*. New York: Jeremy P. Tarcher/Putnam a member of Penguin/Putnam.

Hale, J. 2018. The limitations of science. https://psychcentral.com/blog/the-limitations-of-science/ accessed 6 October 2020.

Hands, J. 2017. *Cosmosapiens: Human Evolution from the Origin of the Universe*. New York: Overlook Duckworth.

Harari, Y. N. 2016. *Sapiens. A Brief History of Humankind*. New York: Signal.
Harari Y. N. 2017. *Homo Deus. A Brief History of Tomorrow*. New York: Signal.
Harari, Y. N. 2018. *21 Lessons for the 21st Century*. New York: Signal.
Harding, L. 2020. COVID-19 panic worse than disease. https://fcpp.org/2020/04/06/covid-19-panic-worse-than-disease/, accessed 4 November 2020.
Harman, W. and Clark, J. (eds.) 1994. *New Metaphysical Foundations of Modern Science*. Sausalito, CA: Institute of Noetic Sciences.
Harman, W. and Sahtouris, E. 1998. Biology Revisioned. Berkeley, CA: North Atlantic Books.
Harris, R. 2017. *Rigor Mortis. How Sloppy Science Creates Worthless Cures, Crushes Hopes, and Wastes Billions*. New York: Basic Books.
Harris, S. 2010. *The Moral Landscape. How Science can determine Human Values*. New York: The Free Press.
Harris, S. 2015. *Waking Up. A Guide to Spirituality without Religion*. New York: Simon & Schuster.
Hawkins, D. R. 2013. *I: Reality and Subjectivity*. West Sedona, AZ: Veritas.
Hayward, J. W. and Varela, F. J. (eds.) 1992. *Gentle Bridges. Conversations with the Dalai Lama on the Sciences of the Mind*. Boston & London: Shambhala.
Hendricks, G., and P. Johncock (eds). 2005. *Already Home. Radiant Wisdom and Life-Changing Meditations from Ramana Maharshi, Sri Nisargadatta, and Teachers of the Advaita Tradition*. Carlsbad, CA: Hay House.
Hesse, H. 1951. *Siddhartha*. Translated by Hilda Rosner. New York: A New Directions Book.
Hillig, C. 2004. The cosmic joke. In: Kersschot, J. *This Is It. Dialogues on the Nature of Oneness*, pp. 163-169. London: Watkins Publishing.
Hoffman, D. 2008. Conscious realism and the mind-body problem. Mind and Matter 6: 87-121.
Hoffman, D. 2018. A universe of consciousness. https://www.youtube.com/watch?v=n6-Q6seVViU, accessed 4 November 2020.
Hoffman, D. 2019. *The Case against Reality. Why Evolution hid the Truth from our Eyes*. New York: WW Norton.
Hoggan, J. 2016. *I'm Right and You're an Idiot. The Toxic State of Public Discourse and how to Clean it up*. Gabriola Island, B.C.: New Society Publishers.
Holdrege, C. 1996. *Genetics and the Manipulation of Life: The Forgotten Factor of Context*. Hudson, NY: Lindisfarne Press.
Hollick, M. 2006. *The Science of Oneness. A Worldview for the Twenty-First Century*. Winchester UK/New York: O-Books.
Hurford, J. R. H. 2014. *The Origin of Language*. Oxford University Press.
Hutchins, G. 2014. *The Illusion of Separation: Exploring the Cause of our Current Crises*. Edinburgh: Floris Books.
Hubbard, R. and Wald, E. 1999. *Exploding the Gene Myth*. Boston, MA: Beacon Press.
Huxley, A. 1978. *The Human Situation*. London: Chatto & Windus.
Izutsu, T. 2012. *Language and Magic. Studies in the Magical Function of Speech*. Kuala Lumpur: The Other Press.
Jabs, H. and Rubik, B. 2019. Detecting subtle energies with a physical sensor array. Cosmos and History: Journal of Natural and Social Philosophy 15 (1).

James, W. 1976. *Essays in Radical Empiricism*. Cambridge, MA: Harvard University Press (originally published in 1912) http://www.informationphilosopher.com/solutions/philosophers/james/Essays_in_Radical_Empiricism.html, accessed 4 November 2020.

Jampolsky, G.G. 1979. *Love is Letting Go of Fear*. Millbrae, CA: Celestial Arts.

Jawer, M. 2021. Sentience, not consciousness, is the key to the cosmos. https://erraticus.co/2020/08/26/sentience-consciousness-cosmos-panpsychism/

Jeune, B. and Sattler, R. 1992. Multivariate analysis in process morphology. Journal of Theoretical Biology 156: 147-167.

Josephson, B. D. and Rubik, B. A. 1992. The challenge of consciousness research. https://www.tcm.phy.cam.ac.uk/~bdj10/mm/articles/athens.pdf, accessed 16 October 2020.

Kabat-Zinn, J. 2011. The healing power of mindfulness. https://www.youtube.com/watch?v=_If4a-gHg_I, accessed 16 October 2020.

Kabat-Zinn, J. 2013. *Full Catastrophe Living. Using the Wisdom of Your Body and Mind to Face Stress, Pain, and Illness*. Revised edition. New York: Bantam/Random House.

Kastrup, B. 2019. Why materialism is a dead end. https://iai.tv/articles/why-materialism-is-a-dead-end-bernardo-kastrup-auid-1271 accessed 16 October 2020

Kastrup, B. 2020. Arthur Schopenhauer: The West's nondual sage. https://www.scienceandnonduality.com/article/arthur-schopenhauer-the-wests-nondual-sage, accessed 15 November 2020.

Katagiri, D. 2007. *Each Moment is the Universe*. Boston: Shambhala.

Katie, B. 2002. *Loving What Is. Four Questions that can Change your Life*. New York: Three Rivers Press.

Katz, R. 2020. A personal response to the "plandemic" claim. http://www.integralworld.net/katz1.html, accessed 25 October 2020.

Kauffman, S. A. 1995. *At Home in the Universe. The Search for Laws of Self-Organization and Complexity*. Oxford/New York: Oxford University Press, also as an audiobook.

Kauffman, S. A. 2010. *Reinventing the Sacred: A New View of Science, Reason, and Religion*. New York: Basic Books.

Keller, E. F. 1983. *A Feeling for the Organism. The Life and Work of Barbara McClintock*. San Francisco, CA: Freeman.

Kellogg, E. W. III 1987. Speaking in E-Prime: An experimental method for integrating general semantics into daily life. Etc.: A Review of General Semantics 44 (2): 118-128. https://www.generalsemantics.org/wp-content/uploads/2011/05/articles/etc/44-2-kellogg.pdfd.

Kellogg, E. W. III and Bourland, D. D. Jr. 1990/91. Working with E-Prime. Some practical Notes. Etc. 47 (4): 376-392. Working_With_E-Prime_Some_Practical_Notes.pdf

Kelly, L. 2015. *Shift into Freedom*. Boulder, CO: Sounds True.

Keltner, D. 2010. Hands-on Research: The Science of Touch. http://greatergood.berkeley.edu/article/item/hands_on_research, accessed 4 November 2020.

Kent, M. 2020. Build a resilient writing practice from the inside out. http://www.sensewriting.org, accessed 4 November 2020.

Kingma, D. R. 1999. *The Future of Love: The Power of the Soul in Intimate Relationships*. New York: Broadway Books.

Kirchoff, B. K. 1995. A holistic aesthetic for science. Journal of Scientific Exploration 9: 565-978.

Kirchoff, B. K. 1996. Scientific communities, objectivity and the transformation of science. Institute of Noetic Sciences: Causality Issues in Contemporary Science, IONS Research Report CP-8, pp. 1-26.

Kirchoff, B. K. 2002. Aspects of Goethean science: complexity and holism in science and art. In: Rowland, H. (ed.), *Goethe, Chaos, and Complexity*. Amsterdam: Editions Rodopi, pp. 79-89, 189-194.

Klinghardt, D. 2020. Corona 2020. https://www.youtube.com/watch?v=NctiERzrny4, accessed 4 November 2020.

Kodish, B. I. 2013. Aristotle's Non-Aristotelianism. Korzybski Files. http://korzybskifiles.blogspot.com/2013/06/aristotles-non-aristotelianism.html, accessed 4 November 2020.

Kodish, S.P. and Holsten, R. P. (eds.). 1998. *Developing Sanity in Human Affairs*. Westport, CT: Praeger.

Kodish, S. P. and Kodish, B. I. 2011. *Drive Yourself Sane*. 3rd edition. Extensional Publishing.

Koestler, A. 1964. *The Act of Creation*. London: Hutchinson.

Korzybski, A. 1958. *Science and Sanity: An Introduction to Non-Aristotelian Systems and General Semantics*. 4th edition. The International Non-Aristotelian Library Publishing Company. (1st ed. 1933; 5th ed. 1994; first CD-Rom ed. 1996: http://esgs.free.fr/uk/art/sands.htm, accessed 4 November 2020.

Korzybski, A, 2010. Selections from *Science and Sanity: In Introduction to Non-Aristotelian Systems and General Semantics*. Second edition. Fort Worth, TX: Institute of General Semantics.

Kosko, B. 1993. *Fuzzy Thinking. The New Science of Fuzzy Logic*. New York: Hyperion.

Kosko, B. 1999. *The Fuzzy Future*. New York: Harmony Books.

Kovacs, B. J. 2019. *Merchants of Light: The Consciousness that is Changing the World*. The Kamlak Center

Kozlovsky, D. G. 1974. *An Ecological and Evolutionary Ethic*. Englewood Cliffs, NJ: Prentice-Hall.

Kripal, J. J. 2019. *The Flip*. New York: Bellevue Literary Press.

Kuhn, T. 2012. *The Structure of Scientific Revolutions*. 50th anniversary edition. Chicago: University of Chicago Press (first published in 1962).

Lakoff, G. and Johnson, M. 2003. *Metaphors We Live By*. Chicago: Chicago University Press.

Lancaster, B. L. 2004. *Approaches to Consciousness. The Marriage of Science & Mysticism*. New York: Palgrave Macmillan.

Lanctôt, G. 1995. *The Medical Mafia. How to get out of it alive and take back our health and wealth*. Miami, FL: Here's the Key.

Lane, D. C. 2012. The Feynman Imperative: Why science works. http://www.integralworld.net/lane35.html, accessed 4 November 2020.

Lane, D. C. and S. Diem-Lane. 2012. The Shiva nature of science. Exploring the multiple forms of gathering knowledge. http://www.integralworld.net/lane38.html, accessed 4 November 2020.

Lao Tzu. 1972. *The Way of Life according to Laotzu* (translation of the *Tao Te Ching* by Witten Bynner). New York: Perigee Books.

Laszlo, E. 1973. A general systems model of the evolution of science. Scientia (Milan) 107: 379-395.

Lent, J. 2017. *The Patterning Instinct. A Cultural History of Humanity's Search for Meaning*. Amherst, NY: Prometheus Books.

Lewontin, R. C. 1991. *Biology as Ideology. The Doctrine of DNA*. Concord, ON: Anansi.

Lipkin, B. H. 2008. *The Biology of Belief: Unleashing the Power of Consciousness, Matter, & Miracles*. Carlsbad, CA: Hay House.

Lorimer, D. 2017. Aim-oriented empiricism. Book review of Maxwell, N. 2017. *Understanding Scientific Progress*. Paradigm Explorer 2017/2, p. 44.

Lorimer, D. 2020a. Heresy, dissidence and authority. Issue 3 of Towards a New Renaissance. Newsletter of the Scientific and Medical Network. https://explore.scimednet.org/wp-content/uploads/2020/06/TANR-Issue-3-FINAL.pdf

Lorimer, D. 2020b. From gain of function to gain of wisdom. Issue 5 of Toward a New Renaissance. Newsletter of the Scientific and Medical Network. https://explore.scimednet.org/wp-content/uploads/2020/08/TANR-issue-5.pdf

Loye, D. 2018. *Rediscovering Darwin. The Rest of Darwin's Theory and why we need it to*day. Glasgow: Romanes Press (for a review see Paradigm Explorer 129 (1), p. 62, 2019).

Lueddeke, G. R. 2016. *Global population health and well-being in the 21st century. Toward new paradigms, policy, and practice*. New York: Springer.

Luisi, P. L. 2009. *Mind and Life. Discussions with the Dalai Lama on the Nature of Reality*. New York: Columbia University Press.

Manek, N. 2019. *Bridging Science and Spirit. The Genius of William A. Tiller's Physics and the Promise of Information Medicine*. Gloucester, UK: Conscious Creation LLC.

Mann, C. 1991. Lynn Margulis: Science's Unruly Earth Mother. Science 252 (5004): 378-3

Marik, P. 2020. EVMC critical care COVID-19 management protocol. MATH+. https://www.evms.edu/media/evms_public/departments/internal_medicine/EVMS_Critical_Care_COVID-19_Protocol.pdf, posted 28 September 2020.

Maxwell, N. 2017. *Understanding Scientific Progress*. St. Paul, MN: Paragon House.

McAllister, J. W. 1999. *Beauty and Revolution*. Ithaca, NY: Cornell University Press.

McFarlane, T. J. 2002. *Einstein and Buddha. The Parallel Sayings*. Berkeley, CA: Ulysses Press.

McLaughlin, C. 2005. Spirituality and ethics in business. https://www.emerald.com/insight/content/doi/10.1108/ebr.2005.05417aab.004/full/html

https://web.archive.org/web/20100430162007/http://www.srmhp.org/0202/pseudoscience.html, accessed 4 November 2020.

McLeish, T. 2019. *The Poetry and Music of Science*. Comparing Creativity in Science and Art. Oxford: Oxford University Press.

McNally, R. J. 2003. Is the pseudoscience concept useful for clinical psychology? The Scientific Review of Mental Health Practice 2 (2) http://www.srmhp.org/0202/pseudoscience.html, accessed 4 November 2020.

Mercola, J. 2020a. *EMF*D. 5G, Wi-Fi & Cell Phones: Hidden Harms and How to Protect Yourself*. Carlsbad, CA: Hay House.

Mercola, J. 2020b. COVID-19 critical care. https://articles.mercola.com/sites/articles/archive/2020/05/29/dr-paul-marik-critical-care.aspx?cid_source=dnl&cid_medium=email&cid_content=art1HL&cid=20200529Z1&et_cid=DM547464&et_rid=882621289, posted 29 May 2020.

Mercola, J. 2020c. Vitamin D combats viral infections and boosts immune system. https://articles.mercola.com/sites/articles/archive/2020/05/31/vitamin-d-combats-viral-infections-boosts-immune-system.aspx?cid_source=dnl&cid_medium=email&cid_content=art1HL&cid=20200531Z1&et_cid=DM554053&et_rid=884087414, posted 21 June 2020.

Mercola, J. 2020d. Undetectable engineering methods used to create SARS-CoV-2. https://articles.mercola.com/sites/articles/archive/2020/06/03/coronavirus-engineered.

aspx?cid_source=dnl&cid_medium=email&cid_content=art1ReadMore&cid=20200603Z1&et_cid=DM554047&et_rid=886176403, posted 3 June 2020.

Mercola, J. 2020e. Peak Internet. The censorship bubble is about to burst. https://articles.mercola.com/sites/articles/archive/2020/06/05/joe-rogan-youtube-censorship.aspx?cid_source=dnl&cid_medium=email&cid_content=art1HL&cid=20200605Z1&et_cid=DM554086&et_rid=887654046, posted 5 June 2020.

Mercola, J. 2020f. Magnesium and K2 optimize your vitamin D supplementation. https://articles.mercola.com/sites/articles/archive/2020/06/15/vitamin-d3-k2-and-magnesium.aspx?cid_source=dnl&cid_medium=email&cid_content=art1ReadMore&cid=20200615Z1&et_cid=DM567263&et_rid=894859166, posted 15 June 2020.

Mintzberg, H. 2015. *Rebalancing Society*. Oakland, CA: Berrett-Koehler Publishers.

Monbiot, G. 2017. *Out of the Wreckage: A New Politics for an Age of Crisis*. New York: Verso.

Mooji. 2017. This exercise is all the help you need. https://www.youtube.com/watch?v=wa5IF7x-ziA, accessed 4 November 2020.

Moromisato, G. 2004. A very short history of humanity. https://www.neurohack.com/earthguide/History.html, accessed 4 November 2020.

Mowrey, D. B. 1998. *The Scientific Validation of Herbal Medicine*. New York: McGraw-Hill.

Mueller, B. 1989. *Goethe's Botanical Writings*. Woodbridge, CT: Ox Bow Press.

Murray, M. et al. 2002. *How to Prevent and Treat Cancer with Natural Medicine*. New York: Riverhead Books.

Newell, K., Kossi, K. and Alexander, E. Sacred Acoustics. Engage Your Infinite Mind. https://www.sacredacoustics.com/pages/our-story accessed 15 October 2020

Nicholson, D. J. and Dupré, J. (eds.) 2018. *Everything Flows. Toward a Processual Philosophy of Biology*. Oxford: Oxford University Press.

Nidamboor, R. 2017. Truth and beauty: The two faces of science. https://www.fairobserver.com/culture/science-research-esthetics-culture-news-19621/, accessed 4 November 2020.

Niebauer, C. 2019. *No Self, No Problem: How Neuropsychology is Catching up to Buddhism*. San Antonio, TZ: Hierophant Publishing.

Nietzsche, F. 2002. *Beyond Good and Evil*. Translated by Judith Norman and edited by Rolf-Peter Horstmann. Cambridge: Cambridge University Press.

Orrell, D. 2012. *Truth or Beauty: Science and the Quest for Order*. New Haven, CT: Yale University Press.

Ortner, N. 2013. *The Tapping Solution*. New York: Hay House (also available as an audiobook).

Oschman, J. L. 2015. *Energy Medicine. The Scientific Basis*. New York: Elsevier.

Osho. 1993. *The Everyday Meditator*. Boston: Charles E. Tuttle.

Osho. 1998. *Take It Really Seriously. A Revolutionary Insight into Jokes*. London: Grace Publishing.

Osho. 1999. *The Secret of Secrets. Talks on the Secret of the Golden Flower*. The Rebel Publishing House.

Osho. 2004. *Meditation. The first and last Freedom*. New York: St. Martin's Griffin.

Osho. 2010. *The Book of Secrets. 112 Meditations to Discover the Mystery Within*. New York: St. Martin's Griffin.

Osho. 2012. *The Heart Sutra: Talks on Sutras of Gautama the Buddha*. Rebel Publishing House.

Osho. 2019. The future of education: five dimensions. https://www.oshotimes.com/insights/society/culture/5-dimensions-of-education/, accessed 4 November 2020

OshoTimes. 2020a. Don't just sit there. https://www.oshotimes.com/opinions/dont-just-sit-there/, accessed 4 November 2020.
OshoTimes. 2020b. The Coronavirus Meditation https://www.oshotimes.com/opinions/the-mind/the-coronavirus-meditation/, accessed 4 November 2020.
Pall, M. L. 2018. 5G: Great risk for EU, U.S. and international health! Compelling evidence for eight distinct types of great harm caused by electromagnetic field (EMF) exposures and the mechanism that causes them. https://peaceinspace.blogs.com/files/5g-emf-hazards--dr-martin-l.-pall--eu-emf2018-6-11us3.pdf, accessed 20 October 2020.
Peitgen, H-O and Richter, P. H. 1986. *The Beauty of Fractals. Images of Complex Dynamical Systems.* Heidelberg/New York: Springer Verlag.
Pinker, S. 2018. *Enlightenment Now: The Case for Reason, Science, Humanism, and Progress.* New York: Viking.
Poonja, H. W. L. 2001. *This. Prose and Poetry of Dancing Emptiness.* New Delhi: Full Circle (Poona is also known as Papaji).
Popper, K. R. 1962. *Conjectures and Refutations. The Growth of Scientific Knowledge.* New York: Basic Books.
Potari, A. D. 2018. The light of Hellenism. https://www.embodiedphilosophy.com/the-light-of-hellenism/, accessed 14 October 2020; also in Paradigm Explorer 132: 12-15, 2020.
Powers, B. 2016. Realizing awakened consciousness. http://www.integralworld.net/powers19.html, accessed 4 November 2020.
Presti, D. E. et al. 2019. *Mind beyond Brain.* New York: Columbia University Press.
Prigogine, I. 1997. *The End of Certainty.* New York: The Free Press.
Quine, W. V. O. 1980. *From a Logical Point of View.* Cambridge, MA: Harvard University Press.
Radin, D. I. 1997. *The Conscious Universe. The Truth of Psychic Phenomena.* New York: HarperEdge of HarperSanFrancisco.
Radin, D. I. 2006. *Entangled Minds. Extrasensory Experiences in a Quantum World.* New York: Pocket Books.
Rajneesh, Bhagwan Shree. 1978. *The Way of Tao.* Delhi: Motilal Banarsidass.
Rajneesh, Bhagwan Shree. 1984. *The Book.* Rajneesh Foundation International.
Rankin, A. 2010. *Many-Sided Wisdom.* Winchester, UK/Washington, USA: O Books.
Ravindra, R. 1991. *Science and Spirit.* New York: Paragon House.
Ravindra, R. 2000. *Science and the Sacred.* Wheaton, IL: The Theosophical Publishing House.
Ravindra, R. 2015-16. Science and Religion. Interview with Ravi Ravindra. Network Review 119: 10-12.
Raza, A. 2019. *The First Cell: And the Human Cost of Pursuing Cancer to the Last.* New York: Basic Books.
Reiss, K. und Bhakdi, S. 2020. *Corona False Alarm?* White River Junction, Hartford, VT: Chelsea Green Publishing.
Reiss, K. and Bhakdi, S. 2021. *Corona Unmasked. New Facts and Figures.* Berlin & Vienna: Goldegg.
Reiss, J. and Sprenger, J. 2014. Scientific Objectivity. Stanford Encyclopedia of Philosophy https://plato.stanford.edu/entries/scientific-objectivity/#VieNow
Risch, H. R. 2020. The key to defeating COVID-19 already exists. https://www.newsweek.com/key-defeating-covid-19-already-exists-we-need-start-using-it-opinion-1519535?amp=1&__twitter_impression=true, posted 23 July 2020.

Rosenberg, M. B. 2003. *Nonviolent Communication: A Language of Life: Life-Changing Tools for Healthy Relationships*. Encinitas, CA: PuddleDancer Press.

Rutishauser, R. 2019. Ever since Darwin. Why plants are important for evo-devo research. In: Fusco, G. (ed.) *Perspectives on Evolutionary and Developmental Biology. Essays for Alessandro Minelli*. Padova: Padova University Press. Chapter 6, pp. 41-56, https://www.padovauniversitypress.it/system/files/attachments_field/9788869381409-oa.pdf

Rutishauser, R. 2020. Evo-devo: past and future of continuum and process plant morphology. Philosophies 5 (4), 41 https://www.mdpi.com/2409-9287/5/4/41/pdf

Rutishauser, R. and Isler, B. 2001. Developmental genetics and morphological evolution of flowering plants, especially bladderworts (Utricularia): Fuzzy Arberian Morphology complements Classical Morphology. Annals of botany 88: 1184-1201. http://www.systbot.uzh.ch/static/personen/rolf_rutishauser_assets/Rut_Isl_AnnBot_2001.pdf

Rutishauser, and Sattler, R. 1985. Complementarity and heuristic value of contrasting models in structural botany. 1. General considerations. Botanische Jahrbücher für Systematik 107: 415-455.

Sabbadini, S. A. 2013. *Lao Tzu: Tao Te Ching. A Guide to the Interpretation of the Foundational book of Taoism*. Published by Shantena Augusto Sabbadini,

Sahtouris, E. 1999. *EarthDance: Living Systems in Evolution*. https://ratical.org/LifeWeb/Erthdnce/erthdnce.html, accessed 4 November 2020.

Sahtouris, E. 2000. *EarthDance: Living Systems in Evolution*. Bloomington, IN: iUniverse.

Sahtouris, E. 2003. After Darwin - reuniting spirituality with science to form a new world view. https://ratical.org/LifeWeb/Articles/AfterDarwin.html, accessed 4 November 2020

Sahtouris, E. 2014. Ecosophy: Nature's guide to a better world. https://www.kosmosjournal.org/article/ecosophy-natures-guide-to-a-better-world/ accessed 2 October 2020.

Sahtouris, E. 2018. *Gaia's Dance: The Story of Earth and Us*. Scotts Valley, CA: CreateSpace.

Sattler, R. 1974. A new approach to gynoecial morphology. Phytomorphology 24: 22-34.

Sattler, R. 1988. A dynamic multidimensional approach to floral morphology. In: Leins, P. et al. (eds.) *Aspects of Floral Development*. Berlin/Stuttgart: J. Cramer, pp. 1-6.

Sattler, R. 1986. *Biophilosophy. Analytic and Holistic Perspectives*. New York: Springer.

Sattler, R. 1992. Process morphology: structural dynamics in development and evolution. Canadian Journal of Botany 70: 708-714. http://www.nrcresearchpress.com/doi/pdf/10.1139/b92-091

Sattler, R. 1994. Homology, homeosis and process morphology in plants. In: Hall, B. K. (ed.) *Homology: The Hierarchical Basis of Comparative Biology*. New York: Academic Press, pp. 423-475.

Sattler, R. 1996. Classical morphology and continuum morphology: opposition and continuum. Annals of Botany 78: 577-581. http://aob.oxfordjournals.org/content/78/5/577.full.pdf

Sattler, R. 1998. On the origin of symmetry, branching and phyllotaxis in land plants. In: Jean, R. V. and Barabé, D. (eds.) *Symmetry in Plants*. Singapore: World Scientific.

Sattler, R. 2001a. Some comments on the morphological, scientific, philosophical and spiritual significance of Agnes Arber's life and work. Annals of Botany 88: 1215-1217.

Sattler, R. 2001b. Non-conventional medicines and holism. Holistic Science and Human Values 5: 1-15.

Sattler, R. 2008. *Wilber's AQAL Map and Beyond*. https://beyondwilber.ca/AQALmap/bookdwl/book_download.html

Sattler, R. 2010. *Healing Thinking and Being.* http://www.beyondwilber.ca/books/healing/healing_thinking_and_being.html

Sattler, R. 2014. Ken Wilber's AQAL Dogma. http://www.beyondwilber.ca/AQALmap/AQAL-dogma/ken-wilber.html

Sattler, R. 2016a. Science and mystery. Holistic Science Journal 3 (1): 49-54.

Sattler, R. 2016b. Wholeness, Fragmentation, and the Unnamable: Holism, Materialism, and Mysticism – a Mandala (this book is continually updated) https://beyondwilber.ca/books/mandala/mandala_of_life_and_living.html

Sattler, R. 2017. Morphological development of flowers. https://www.beyondWilber.ca/about/flower/floral-development.html

Sattler, R. 2018. Philosophy of plant morphology. Elemente der Naturwissenschaft 108: 55-79. https://elementedernaturwissenschaft.org/en/node/1401 and https://beyondwilber.ca/about/plant-morphology/philosophy-of-plant-morphology.html

Sattler, R. 2019. Structural and dynamic approaches to the development and evolution of plant form. In: Fusco, G. (ed.) Perspectives on Evolutionary and Developmental Biology. Essays for Alessandro Minelli. Padova: Padova University Press, Chapter 6, pp. 57-70. https://www.padovauniversitypress.it/system/files/attachments_field/9788869381409-oa.pdf

Sattler, R. and Jeune, B. 1992. Multivariate analysis confirms the continuum view of plant form. Annals of Botany 69: 249-262.

Sattler, R. and Rutishauser, R. 1990. Structural and dynamic descriptions of the development of *Utricularia foliosa* and *U. australis*. Canadian Journal of Botany 68: 1989-2003.

Scales, S. 1995. Values in Ethics and Science: A Case Against Objective Moral Realism. Dissertation. University of California, San Diego. https://philpapers.org/rec/SCAVIE, accessed 4 November 2020.

Scales, S. 2002. Value-ladenness, theoretical virtues, and moral wisdom. Teaching Ethics 2 (2): 19-28. https://www.academia.edu/15043132/Value_ladenness_Theortetical_Virtues_and_Moral_Wisdom, accessed 6 January 2020.

Schick N. 2020. *Deepfakes: The Coming Infocalypse.* New York: Twelve.

Schneider, K. 2013. *The Polarized Mind: Why it's killing us and what we can do about it.* Colorado Springs, CO: University Professors Press.

Schroeder, M. J. 2016. The philosophy of philosophies. Synthesis through diversity. Philosophies 1: 68-72; https://www.mdpi.com/2409-9287/1/1/68/htm, accessed 19 October 2020.

Schwartz, S. A., Woollacott, M. and Schwartz, G. E. (eds.) 2020. *Is Consciousness Primary?* The Academy for the Advancement of Postmaterialist Sciences.

Sheldrake, R. 2012/20. *The Science Delusion. Freeing the Spirit of Enquiry.* London: Coronet (also published as *Science Set Free*. New York: Deepak Chopra Books.

Shepherd, L. J. 1993. *Lifting the Veil. The Feminine Face of Science*. Boston & London: Shambhala.

Shiva, V. 2020. One planet, one health – connected through biodiversity: from the forests, to our farms, to our gut microbiome. https://www.navdanya.org/bija-reflections/2020/03/18/ecological-reflections-on-the-corona-virus/, accessed 20 October 2020.

Siegfried, T. 2019. Could quantum mechanics explain the existence of space-time? https://astronomy.com/news/2019/05/could-quantum-mechanics-explain-the-existence-of-space-time, posted 7 May 2019.

Somers, S. 2009. *Knockout: Interviews with doctors who are curing cancer - and how to prevent getting it in the first place*. New York: Harmony Books.

Spielmann, R. 1998. *'You're so Fat!' Exploring Ojibwe Discourse*. University of Toronto Press.

Spiro, R. 2015. The nature of consciousness. https://www.youtube.com/watch?v=R-IIzAblVlg accessed 16 October 2020.

Steiner, R. 1918. *A Road to Self-Knowledge*. New York: G. P. Putnam's Sons (originally published in German, 1912).

Steiner, R. 1986. *At the Gates of Spiritual Science*. Hillside, UK: Rudolf Steiner Press.

Steiner, R. 2002. *Occult Science – An Outline*. New York: Anthroposophic Press.

Stockdale, S. 2009. Here's Something About General Semantics. A Primer for Making Sense of Your World. Santa Fee, NM: Published by Steve Stockdale. https://thisisnotthat.com/tintdocs/HSGS-CN-course.pdf

Stockdale, S. 2020. The Structural Differential https://www.thisisnotthat.com/structural-differential/, accessed 4 November 2020.

Stone, D., Patton, B., and S. Heen. 2010. *Difficult Conversations. How to discuss what matters most*. New York: Penguin Books.

Straine, G. (ed.). 2017. *Are there Limits to Science?* Newcastle upon Tyne, UK: Cambridge Scholars.

Stromberg, J. 2015. The hygiene hypothesis: How being too clean might be making us sick. https://www.vox.com/2014/6/25/5837892/is-being-too-clean-making-us-sick, accessed 4 November 2020.

Tanahashi, K. 2014. *The Heart Sutra*. Boston: Shambhala.

Tannen, D. 2001. *You Just Don't Understand. Women and Men in Conversation*. New York: William Morrow and Co.

Tannen, D. 2011. *That's Not What I Meant. How Conversational Style Makes or Breaks Relationships*. New York: HarperCollins.

Tannen, D. 2015. *Handbook of Discourse Analysis*. 2nd ed. New York: Wiley.

Tao Te Ching, translated by Stephen Mitchell. 1992. New York: HarperPerennial.

Tart, C. T. 1972. States of consciousness and state-specific sciences. Science 176: 1203-1210.

Taylor, E. 1994. Radical empiricism and the conduct of research. In: Harman, W. and Clark, J. (eds.) *New Metaphysical Foundations of Science*. Sausalito, CA: Institute of Noetic Sciences, pp. 345-374.

Taylor, S. 2018. *Spiritual Science: Why Science needs Spirituality to Make Sense of the World*. London: Watkins.

Teicholz, N. 2015. *The Big Fat Surprise*. New York: Simon & Schuster.

Thibodeau, M. 2016. *Immune System Boosters: How to Naturally Boost Your Immune System & Stay Healthy All Year Long*. Hubbardston, MA: Boondocks Botanicals.

Thich Nhat Hanh. 1988. *The Heart of Understanding. Commentaries on the Prajnaparamita Heart Sutra*. Berkeley, CA: Parallax Press.

Thich Nhat Hanh. 2003. *No Death, no Fear: Comforting Wisdom for Life*. New York: Riverhead Books.

Thich Nhat Hanh. 2014a. *How to Love*. Berkeley, CA: Parallax Press (also available as an e-book).

Thich Nhat Hanh. 2014b. Interessere interbeing. https://www.youtube.com/watch?v=kgU8sLTZduQ, accessed 4 November 2020.

Tiller, W. A. 1997. *Science and Human Transformation. Subtle Energies, Intentionality and Consciousness*. Walnut Creek, CA: Pavior.

Tiller, W. A. 2007. *Psychoenergetic Science: A Second Copernican-Scale Revolution.* Walnut Creek, CA: Pavior.

Tolle, E. 2004. *The Power of Now. A Guide to Spiritual Enlightenment.* Vancouver: Namaste Publishing and Novato, CA: New World Library.

Tudge, C. 2021. *The Great Re-Think: A 21st Century Renaissance.* Pari, Italy: Pari Publishing.

Vita, E. 2017. *Instant Presence: Allow Natural Meditation to Happen.* London: Watkins.

Wagner, A. 2015. *The Arrival of the Fittest: How Nature Innovates.* London: Current.

Wahl, D. C. 2017. "Zarte Empirie": Goethean science as a way of knowing. https://medium.com/age-of-awareness/zarte-empirie-goethean-science-as-a-way-of-knowing-e1ab7ad63f46, accessed 4 November 2020.

Walach, H. 2015. *Secular Spirituality.* New York: Springer.

Walach, H. 2018. Report of the Galileo Commission Project. Paradigm Explorer 128: 5-10.

Wallace, B. A. 2000. *The Taboo of Subjectivity: Toward a New Science of Consciousness.* New York: Oxford University Press.

Waller, J. 2004. *Fabulous Science: Fact and Fiction in the History of Scientific Discovery.* Oxford: Oxford University Press.

Wangyal, Tenzin, Rinpoche. 2011. *Tibetan Yogas of Body, Speech, and Mind.* Boston & London: Snow Lion.

Warzel, C. 2018 (February 11). He predicted the 2016 fake news crisis. Now he is worried about an information apocalypse. https://www.buzzfeednews.com/article/charliewarzel/the-terrifying-future-of-fake-news, accessed 4 November 2020.

Watts, A. 1951. *The Wisdom of Insecurity.* New York: Vintage Books.

Watts, A. 1957. *The Way of Zen.* New York: Vintage Books.

Watts, A. 1975. *Tao: The Watercourse Way.* New York; Pantheon Books.

Weil, A. 2004. *Health and Healing: The Philosophy of Integrative Medicine and Optimum Health.* New York: Mariner Books. Revised edition.

Weil, A. 2009. *Why our Health matters. A Vision of Medicine that can transform our Future.* New York: Hudson Street Press.

Weil, A.T. 2011. *Spontaneous Happiness. A New Path to Emotional Well-Being.* New York: Little, Brown & Co.

White, L. 1967. The historical roots of our ecological crisis. Science 155: 1203-1207.

Whitehead, A. N 1929. *Process and Reality. An Essay in Cosmology.* New York: Macmillan.

Whorf, B. 2012. *Language, Thought, and Reality. Selected Writings by Benjamin Lee Whorf* (edited by J. B. Carroll et al.). Cambridge, MA: MIT Press.

Wilber, K. 1998. *The Marriage of Sense and Soul: Integrating Science and Religion.* New York: Random House.

Wilber, K. 1999. *The Spectrum of Consciousness* (originally published in 1977). In: The Collected Works of Ken Wilber. Vol. 1. Boston & London: Shambhala.

Wilber, K. 2000. *Sex, Ecology, Spirituality.* 2nd edition. Boston & London: Shambhala (1st edition 1995).

Wilber, K. 2001. *A Theory of Everything.* Boston: Shambhala.

Wilber, K. 2006a. Excerpt G: Toward a comprehensive theory of subtle energies. http://www.kenwilber.com/writings/read_pdf/87

Wilber, K. 2006b. *Integral Spirituality.* Boston: Integral Books.

Wilber, K. 2006c. Ken Wilber stops his brainwaves. https://www.youtube.com/watch?v=LFFMtq5g8N4, accessed 4 November 2020.

Wilber, K. 2017. *The Religion of Tomorrow*. Boulder, CO: Shambhala.

Wilber, K. et al. 2008. *Integral Life Practice: A 21st-Century Blueprint for Physical Health, Emotional Balance, Mental Clarity, and Spiritual Awakening*. Boston & London: Shambhala.

Wildwood, R. 2018. *Primal Awareness*. Abingdon, Oxon, UK: Moon Books (John Hunt).

Wilhelm, R. 1962 Translator of *The Secret of the Golden Flower. A Chinese Book of Life*. Foreword by C. G. Jung. London: Routledge & Kegan Paul.

Wiseman, R. and Schlitz, M. 2019. Experimenter effect and the remote detection of staring. https://www.researchgate.net/publication/238231060_Experimenter_effects_and_the_remote_detection_of_staring, accessed 4 November 2020.

Wolfe, G. W. 2014. *Meditations on Mystery. Science, Paradox, and Contemplative Spirituality*. Lake Oswego, OR: Dignity Press.

Woodger, J. H. 1967. *Biological Principles. A Critical Study*. London: Routledge and Kegan Paul (first edition 1929).

Young, Shinzen. 1997. *The Science of Enlightenment*. Audiotapes. Boulder, CO: Sounds True.

Young, Shinzen. 2011-2016. *Five Ways to Know Yourself. An Introduction to Basic Mindfulness*. eBook. https://www.shinzen.org/wp-content/uploads/2016/08/FiveWaystoKnowYourself_ver1.6.pdf

Young, Shinzen. 2016. *The Science of Enlightenment. How Meditation Works*. Boulder, CO: Sounds True (book).

Zeilinger, A. 2016. Quantum entanglement is independent of space and time. https://www.edge.org/response-detail/26790), accessed 18 December 2020.

About the Author

Photo by Rory Skelly

Rolf Sattler, PhD, DSc (h. c.) FLS, FRSC, Professor Emeritus of McGill University in Montreal, Canada was born in southern Germany. After studying botany, zoology, chemistry, pedagogy, and philosophy at universities in Germany, Austria, and Switzerland, he received his doctorate, with summa cum laude, in natural science from the University of Munich. Subsequently he immigrated to Canada where he became professor at McGill University. He devoted his professional life to the study plant morphology, the development and evolution of plant form. His fascination with the forms and beauty of plants is beyond description. In addition, the investigation of the theoretical and philosophical foundations of plant morphology and science in general were central to his research and his teaching (see Appendix 1). Other explorations included a holistic approach to health and sanity (see Appendix 2) and the human condition and its transcendence (Appendix 3). He published nearly a hundred research papers and several books, including the *Organogenesis of Flowers. A Photographic Text-Atlas* (1973) and *Biophilosophy. Analytic and Holistic Perspectives* (1986). His latest book *Wholeness, Fragmentation, and the Unnamable* (that he published on this website) presents a synthesis of holistic science, materialist mainstream science, and the Unnamable. Much of his research was enriched immensely through collaboration with graduate students, postdoctoral fellows and research associates. He taught courses in plant biology, general biology, the history and philosophy of biology, biology and the human condition, and he gave lectures at many institutions and universities around the world, including Harvard University and the Universities of California, Berlin, Zürich, Delhi, Malaya, and Singapore. His publications are read in 33 countries.

Besides a great interest in the arts, he explored the relation of science and spirituality. At Naropa Institute (which became Naropa University) he taught a summer course on Modern Biology and Zen, and for the Dalai Lama's 60th Birthday Celebrations he was invited to give a lecture on Life Science and Spirituality in a symposium on Divergence and Convergence of Sciences and Spirituality.

He received many honours and awards, including an honorary doctorate (DSc h. c.) from the Open International University at Colombo, Sri Lanka, for his contributions to alternative holistic medicine. He is a fellow of the Linnean Society of London (FLS) and the Royal Society of Canada (FRSC). For more see his website www.beyondWilber.ca.

He can be reached through the contact link on his website www.beyondWilber.ca

Index

A

Abstraction, 30-35, 92
Acceptance, 8, 41, 100, 101, 118-119, 129, 131
Ad hoc hypotheses, 41, 85
Alexander, Eben, 52
Alford, Dan "Moonhawk," 36
All, Leyla, 5, 43
Alternative medicine, 4, 71, 98-100
Amerindian languages, 6
Anecdotal evidence, 45
Angell, Ian O., 7
Animal ancestry, our, 110-112
AQAL map, 56, 81, 93, 94, 106
Aquinas, Thomas, 121
Arber, Agnes, V, X, 6, 21, 78, 91, 92, 95, 115
Arendt, Hannah, 107, 123
Aristotle, 21, 121-123
Aristotelian logic, 21-22, 89, 117, 132
Artificial intelligence (AI), 132
Art, XIII, XV, 30, 38, 73-75, 76, 80, 117
Attar of Nishapur, 81
Attractor, 2
   strange, 2
Auras, 15
Aurobindo, 26

B

Balance, III, XIV, 54, 66, 70, 83, 90, 96, 104, 121, 124, 132
Barke, Jeffrey I, 64
Beauregard, Mario, 48, 52
Beauty, 73-74
Beauty and truth, 74
Belief perseverance, 17
Berendt, Joachim-Ernst, 6, 105
Bhakdi, Sucharit, 60
Bias, 16-18
   confirmation, 17
   consensus, 18

publication, 18
reiteration, 18
selection, 17
Big Three, 56
Biomedical research, 63
   dangers, 63
Bio-psycho-socio-spiritual dimensions, 70
Blackstone, Judith, 50, 53, 99, 127, 128, 130
Blake, William, 40, 47, 90
Bodian, Stephan, 55, 63, 107
Body, 97, 118, 128
   3 bodies, 97
   causal, 97
   emotional, 97
   physical, 97
   subtle, 97
   universal, 97
   very subtle, 97
   vital, 97
Bohm, David, XVI, 35, 58
Bohr, Niels, 20, 25
Bonobo behaviour, 111
Borges, Jorge Luis, 6
Bortoft, Henri, 7
Bos, Arie, 69
Both/and logic, 22-25, 91, 115
Boundlessness, 78
Bowden, John, 62
Brach, Tara, 8, 119, 129, 130
Brain, 99
   heart, 99
   second, 99
   three brains, 99
Brennan, Barbara Ann, 15, 97
Brooks, J. S. 103
Buddha, 79, 81, 95
Buddha's Flower Sermon, 79
Buddhist logic, 23-24
Büller, Harry R., 11
Burch, Vidyamala, 118
Bush, Zach, 62
Butterfly effect, 2

C

Caddy, Eileen, 14
Cage, John, 79
Callison, Candis, 4
Campbell, Joseph, 123, 125, 130
Capitalism, 49, 63, 132
Capra, Fritjof, 77, 96, 107
Chagall, Marc, 75
Chaos theory, 2
Chase, Stuart, 102
Cheng, Richard Z., 60
Chimpanzee behaviour, 111-112
Chinese language, 36
Chopra, Deepak, XIV, 14, 21, 27, 46, 50, 58, 97, 100, 131
Chua, Amy, 111
Church, Dawson, 16, 18, 49, 97
Clear light, 100
Communication, 101-104
   non-violent, 104
Compassion, XI, 36, 67, 78, 98, 108, 112, 116, 119, 127-129
Complementarity, 25, 89, 91, 115
Confirmation bias, 4
Confirmation holism, 42
Conflicts, 119
Consciousness, 52-53
   fundamental, 53
   the hard problem, 52
   individual, 53
   primacy of, 52
   state of, 15
   universal, 53
Conspiracy theory, 44, 62
Consumerism, 49
Contemplation, XV, 2, 27, 70
Context, 10, 49, 57
Contraries, 92
   Coincidence of, 92, 115
Cousins, Norman, 45
COVID-19 pandemic, IX, 18,

43, 44, 59-64, 71-72
 prevention, 61-63
 treatment, 60-63
Cox, Christoph, 7,
Csikszentmihalyi, Mihaly, 12
Culture, 59-72
Curie, Marie, 75
Cushnir, Raphael, 100
Cusset, Gérard, 17, 90, 91

D
Dalai Lama, V, XI, XIII, 12, 21, 25/26, 51, 52, 53, 56, 58, 67, 68, 78, 81, 82, 83, 132
Darwin, Charles, 65
Darwinitis, 69
Dataism, 126
Davidson, Richard J., 18, 55, 132
Davis, Iain, 62
Dawes, Milton, 6
Day, Christopher, 118
Delusion, XIV, all chapters, 87
Dennett, Daniel, 2, 9
Deutscher, Guy, 36
De Waal, Frans, XVI, 66, 67, 111
Diamond, Jared, 111
Diem-Lane, Andrea, 24, 68
Dirac, Paul, 74
Disproof, 6
 no, 6-7
Dogmas, 50-51, 81
Doidge, Norman, 69, 99
Dossey, Larry, 51, 100
Double-blind experiments, 10-12
Dupré, John, 36, 49

E
Ecological and evolutionary ethics, 66
Ecological crisis, 49
Ecology, 50, 63, 78
 deep, 50
Economy, X, XIII, 50, 60, 63, 64
Ecosophy, X, 50
Education, 109, 112

 5 dimensions of, 109
Edwards, Janet, 4, 43
Edwards, Paul, 21, 22
Einstein, Albert, X, 29, 47, 74, 75, 83, 132
Eisenstein, Charles, 42, 44, 60, 62, 126, 131
Either/or logic, 21-23, 89, 116-118
Ellis, Albert, 6
Emergent properties, 49-50
Emptiness (in Buddhism), 78
Emotion, 118
Emotional intelligence, 100
Empiricism, 38- 47, 87
 delicate, 39
 radical, 39
Energies, 50, 97
 gross (physical), 50
 subtle, 50
 very subtle (causal), 50
 continuum of, 50
Epigenetic medicine, 49, 59
Environment, 49, 62-63, 132
Epiviral medicine, 59
Epstein, Mark, 99, 100
Evolutionary theory, 65-66
Etc., 33-34
Ethics, XIII-XIV, 66-67
 naturalistic, 66
Excluded middle, 21-24
Expectancy effect, 16
Experience, 38-39, 46
 direct, 39
 inner, 39
 mental, 39
 pure, 39
 outer, 39
 religious, 38
 sensory, 39
 spiritual, 38
Experimenter effect, 16
Experiments, 10
 controlled, 10
 randomized, 10
 double-blind, 10
Explanation, 1
Extensional devices, 33-34, 101-103

F
Facts, 6, 7, 38
 theory-laden, 6
 value-laden, 7
Falconar, Ted, 29, 30, 101, 113, 124
Falsification, 6
 no, 6
Fascia, 99
Feminine side of science, 53-54
Ferguson, Niall, 126
Ferrer, Jorge N., 19, 77, 80
Feuerstein, Georg, 129
Feyerabend, Paul, 1, 68-70, 84
Feynman, Richard, 8
First person and third person, 8, 78-79, 82
Firstenberg, Arthur, 62
Flow, 12, 35-36, 92
Flowers, 91
Folk knowledge, 12
Folk medicine, 12
Foster, Jeff, 118, 119, 130
Foucault, Michel, 43, 124
Fractals, 74
Franklin, Rosalind, 74
Freedom, XV, 40, 72, 87, 107, 119, 127, 130
 absolute, 107
 health, 63
Fromm, Erich, 107
Fuzzy logic, 26-27, 89-90

G
Gabriel, Markus, 69
Gaille, Louise, 11
Galileo Commission Project, 51-52
Gelman, Andrew, 19
General Semantics, 29, 34, 124
Genesis in Bible, 1
Gerber, Richard, 97
Gerretsen, Isabelle, 62
Giere, Ronald N., 45, 65
Glazier, Jacob W., 19
Glynn, Ian, 74
Gober, Mark, 52
God, 67-68
Goethe, Johann Wolfgang von,

7, 17, 115
Goethean science, 7, 39
Goleman, Daniel, 18, 55, 100, 108, 132
Goldacre, Ben, 10
Goldberg, Jonah, 111
Goodall, Jane, 54
Goodwin, Brian, 50
Gorbachev, Mikhail, 49
Gould, Stephan Jay, 75
Governments, 43, 61, 64-65, 71-72
Gray, John, 104
Greene, Debra, 50, 67, 97
Griffin, David, Ray, 54
Grof, Stanislav, 52, 80
Growth, 126
 continued, 126
 obsession with, 132
Gurdjieff, George, 81, 127

H
Hagemann, Wolfgang, 91
Hale, Jamie, 4
Hands, John, 18, 66, 119, 120
Hanson, Norwood Russell, 7
Happiness, XIII, XIV, XVI, 7, 12, 19, 23, 31, 87, 100, 108, 117, 131
 independent of conditions, XIV
Harari, Yuval Noah, 70, 111, 114, 115, 119, 124, 125, 126, 132
Harman, Willis, 50
Harris, Richard, 10, 11
Harris, Sam, XIII, 66
Hawking, Stephen, 70, 132
Hawkins, David R., 19
Hayward, Jeremy W., XI
Health, XIII, XIV, 23, 26, 33, 49, 51, 60-62, 69, 72, 83, 87, 90, 96-109, 121, 132
Healing logic, 23, 27
Healing sounds, 104
Healing thinking, 27
Healthy language, 33-35
Healthy language-behaviour, 101-104

Healthy lifestyle, 97-98
Healthy thinking skills, 106-107
Heart Sutra, 78, 83, 87
Heisenberg, Werner, 8, 36, 58
Henderson, Sakej, 36
Hendricks, Gay, A2
Heraclitus, 19, 81, 92
Hesse, Hermann, 23, 115
Hierarchy, 93, 94
Hoffman, Donald, 52, 125
Hoggan, James, 27, 118
Holarchy, 93, 94
Holdrege, Craig, 49
Holiness, 90, 96
Holistic science, 49-58, 59-64, 77-78, 85
Hollick, Malcolm, 21, 78, 116
Holomovement, 37
Hubbard, Ruth, 49
Human acquisitions, 112-126
 ideas, 114
 language, 113, 119
 large groups, 112
 logic, 116
 myths, 125-126
 primeval thinking, 120
 philosophical thinking, 120
 scientific thinking, 123
Human condition, 110-132
Human evolution, 119-124
Humanity, 131
 future of, 131-132
Humour, 101
Hurford, James R., 36
Hutchins, Giles, 114
Huxley, Aldous, 114

I
Ideas, 114-116
 noble, 115
Identification, 127
 is the only sin, 127
Identity, 21
 cult of, 21
 myth of, 33
Imbalances, 107
 in society, 107-108
Indescribable, 21, 24, 90

Inergy continuum, 50
Infinite, 40
Infinity, IX
Info apocalypse (infocalypse), 126
Insanity, XIV, all chapters, 87
Insight, X, XV, 19, 29, 55, 67, 70-71, 78, 80, 84, 87, 81-82, 92, 106, 111, 113, 115, 116, 119, 120, 122, 129, 130
 personal, XIV, 3, 9, 80-81, 84
 beyond science, 3
 beyond space and time, 9
Integral vision, 106
Inter-being, 78
Inter-subjectivity, 15
Interconnectedness, 78
Izutsu, Toshihiko, 113

J
Jabs, Harry, 50
Jain(a) logic, 5, A1
James, William, 39, 58, 126
Jampolsky, Gerald, 33
Jawer, Michael, 53
Jeune, Bernard, 89, 90, 91, 95
Jokes, XV
Josephson, Brian, 55

K
Kabat-Zinn, Jon, 106
Kant, Immanuel, 39
Kastrup, Bernardo, 52
Katagiri, Dainin Roshi, 107, A3
Katie, Byron, 130
Katz, Richard, 63
Kauffman, Stuart, 68
Keller, Evelyn Fox, 53
Kellogg, E. W., 34
Kelly, Loch, 2, 40, 119, 127, 130
Keltner, Dacher, A2
Kennedy, Robert F., 63
Kent, Madelyn, 40
Kierkegaard, Soren, 33
Kingma, Daphne Rose, 120
Kirchoff, Bruce K., 15, 18, 50, 54
Klinghardt, Dietrich, 60

Kodish, Bruce, 34, 101, 123
Koestler, Arthur, 73
Korzybski, Alfred, V, X, 8, 29-35, 37, 80, 92, 101, 104, 113, 116, 117, 124, 128
Kosko, Bart, 26, 42, 47
Kovacs, Betty J, 120
Kozlovsky, Daniel G, 66
Kripal, Jeffrey J., 52, 125
Krishnamurti, Jiddu, 109
Kuhn, Thomas, 41

L
Lakoff, George, 7, 53
Lancaster, Brian L., 55
Lanctôt, Guylaine, 43, 98
Lane, David C., 68
Language, 29- 37, 87
 harmful and healthy, 33
Lao Tzu or Laozi, 19, 20, 81, 115-116
Laughter, 94
 yoga, 94
 of the Dalai Lama, 94
Laws as habits, 51
Laws of thought, 21
 beyond, 21
Leaf, 6-7
 partial-shoot theory of the, 6-7
Lent, Jeremy, 121, 123
Lewontin, Richard, C. ("Dick"), 49, 59, 65, 66, 69
Lifestyle, 60, 61, 97
Lightness, XV,
Limitations of science, XIII, all chapters, 84-88
Lipkin, Bruce, 49
Logic, 21-28, 116-118
 Aristotelian, 21, 89, 117
 both/and, 25, 89, 91
 Buddhist, 23, 89
 either/or, 21, 89
 healing, 27, 89
 fuzzy, 26, 89, 91
 Jain(a), 24, 89
Lorimer, David, 44, 62, 63, 70
Love, X, 12, 53-54, 67, 81, 87, 108, 121, 129

romantic, 129
Loye, David, 65
Lueddeke, George R., 98
Luisi, Pier, Luigi, 52
Lu-Tsu, 16

M
Machine metaphor, 48
Magritte, René, XV, 30, 93, 117
Mahakashyap, 79
Maitri Upanishad, 28
Mandala, 56-57
Mandalic view, 56-57, 94
Manek, Nisha, 50
Map is not the territory, 32, 92, 117
Margulis, Lynn, 66
Materialism, 48-49, 125
Mathematics, 125
Matter, 50, 51, 53, 77
Maxwell, Nicholas, 70
McAllister, James W., 74
McClintock, Barbara, 53-54, 90
McFarlane, Thomas, 26, 47, 77, 78
McLaughlin, Corinne, 107
McLeish, Tom, 73
McLuhan, Marshall, 113
McNally, Richard J., 44
Mechanism, 48
Mechanistic materialist science, 48-49, 71
Medawar, Peter, 18
Medical establishment, 43, 61, 71
Medical-political complex, 63
Medicine, 4
 alternative, 4, 5, 51, 98
 conventional, 4
 energy, 98
 epigenetic, 59
 holistic, 59, 61, 62-63, 98
 indigenous, 12
 integrative, 60, 61, 98
 mechanistic, 48, 51
 traditional, 12
 traditional Chinese, 61
Meditation, XV, 8, 13, 19, 37, 39, 40, 47, 57, 60, 75, 80, 82, 87

Melucci, Alberto, 37
Mercola, Joseph, 60-63
Merrell-Wolff, Franklin, 37
Merxmüller, Hermann, 95
Metahuman, 8
Microbial world, 62
Militarism, 49
Mind, 105
 beyond the brain, 51
 cosmic (universal), 107
 divine (undeluded), 107
 thinking, 105
 witnessing, 106
Mintzberg, Henry, 107
Misconceptions about science, IX, XIII, all chapters, 84-85
Monbiot, George, 126
Morals, 66-67
Mosuo tribe, 120
Mueller, Bertha, 18
Murray, Michael, 98
Mysterious, X, 31, 90
Mystery, IX, 3, 77, 79, 87, 94
Myths, 125-126

N
Nada Brahma, 104-105
Nagarjuna, 23
Near-death experiences (NDEs), 52
Neither/nor, 23
Neo-Darwinism, 65
Neti neti, 23
Network thinking, 94
Neuromania, 69
Neuroplasticity, 69, 98-99
Newell, Karen, 46
Newton, Isaac, 7
Nicholson, Daniel J., 36
Nidamboor, Rajgopal, 74
Niebauer, Chris, 30
Nietzsche, Friedrich, 7, 115
Nisargadatta, 81, 97, 128
Non-Aristotelian Systems, 34
Non-identity, 21-23, 29-32, 92-93
Non-local perception, 52
Nonduality, XIV, XV, 19, 67, 77, 78, 88, 121

Nootka language, 36
No-thingness, 78
Noun-verb structure, 35
Nouwen, Henri J. M., 7

O
Objectivity, 15-20, 84
Observer effect, 16
Oneness, XIV, XV, 3, 8, 19, 35, 36, 40, 77-79, 89, 90, 116, 123, 129
Ontological relativity, 42
Orrell, David, 74
Ortner, Nick, 98
Oschman, James L, 98, 98
Osho, IX, X, XV, 14, 16, 28, 37, 55, 57, 79, 80, 82, 104, 105, 109, 113, 130

P
Panzer, Ulrich, VIII, X, 76
Parapsychology, 52
Pall, Martin L., 62
Pattanai, Devdutt, 20
Peace, XIII, XIV, 7, 14, 19, 25, 27-28, 31, 36, 49, 66, 72, 87, 106, 108, 111, 115, 118, 127-129, 131
Peer review, 43
Peitgen, Heinz-Otto, 74
Perspectivism, 24
Pharmaceutical industry, 63, 64, 71-72
Pinker, Steven, 122
Placebo, 11
Planck, Max, 41, 53
Plant morphology, 89-95
  classical, 89
  continuum, 89
  process, 91-92
Plato, 40, 121-123
Platonism, 121
Politics, X, XIII, 21, 26, 43, 63, 91
Pollution, 62-63
Poonja, H. W. L. (Papaji), A3
Popper, Karl, 6, 17
Potari, Athena Despoina, 40, 46, 121, 123

Power drive, 42-43, 111-112, 124-125, 132
Power-knowledge, 42-43, 85, 124-125
Power of science, 1-3
Powers, Barclay, 97
Prediction, 1
Presti, David E., 52
Previous lives, 52
Prigogine, Ilya, 8
Primitive traits, 111
  amplification of, 119
Probabilistic prediction, 1
Process language, 35-37
Process morphology, 91
Process philosophy, 92
Proof, 4
  no, 4-5, 84
Pseudoscience, 44
Psychic phenomena, 52
Psycho-logical, 35
Psycho-neuro-immunology, 69
Publication bias, 18
Pure experience, 39

Q
Quantum physics, 7-8, 35-36, 77-78,
Quine, Willard Van Orman, 42

R
Radical empiricism, 39
Radical view of science, 68-70
Rajneesh, Bagwan Shree, 107, 114
Ramana Maharshi, 83, 105
Rankin, Aidan, 24
Ravindra, Ravi, V, X, 37, 56, 68, 90
Raza, Azra, 50, 98
Reality, III, XIV, XVI, 1, 6, 8, 9, 13, 15, 16, 18-20, 21, 23-26, 29-39, 38-40, 42, 46-49, 51-52, 56-57, 58, 73, 75, 77-78, 81, 84-87, 91, 93, 06, 99, 101-105, 113-114, 116-117, 121, 124-127, 132
Reiss, Julian, 18
Reiss, Karina, 60

Reiteration bias, 18
Relaxation, 40, 60, 101
Religio, 79
Religion, 67
  beyond, 80
Religious experience, 67, 79, 125
Replicability, 10-12, 84
Replication, 10-12, 84
Rheomode, 35
Risch, Harvey A, 60, 61
Rogers, Fred, 14
Rosenberg, Marshall B., 104, 108
Rubik, Beverly, 62
Rumi, 3, 81, 116, 129
Rutishauser, Rolf, 25, 65, 89, 90, 91, 95

S
Sabbadini, Shantena Augusto, III, 75, 80, 114
Sacredness, 90
Sahtouris, Elisabet, V, 50, 53, 54, 66, 78
Sanity, XIV, all chapters, 87, 96-109
Sattler, Rolf, IX, 1, 12, 15, 17, 25, 27, 30, 31, 56-57, 68, 81, 89-95, 106
Scales, S., 7
Schick, Nina, 126
Schlitz, Marilyn, 16
School of Athens, 122
Schopenhauer, Arthur, 41, 53, 105
Schrödinger, Erwin, 53
Schroeder, Marcin J., 69
Schwartz, Gary E., 53
Science, all chapters
  as belief system, 50
  broad, 39, 54-55
  dangers of, 70
  delusion in, 50
  of enlightenment, 55
  feminine side of, 53-54
  holistic, 49-57, 85
  limitations of, IX, all chapters, 84

materialist, 48-49
mechanistic, 48-49, 85
method, 1, 68
misconceptions about, IX, 84-85
narrow, 39
of oneness, 77-78
power in, 42-43, 124-125
power of, 1-3
pseudo, 44, 72
as (pseudo)religion, 69, 77
radical view of, 68-70
sacred, 56
set free, 50-51
simplifications in, 42
spiritual, 56
state-specific, 16
Science and the arts, 73-75
Science and ethics, 66-67
Science and morals, 66
Science and politics, 61-64
Science and religion, 67
Science and the Sacred, X, 56
Science and society, 64-66
Science and Spirit, X, 56
Science and spirituality, 77-83
convergence of, 77-78
difference between, 77-82
divergence of, 78-82
Scientific and Medical Network, 52
Scientific laws, 1
Scientific method, 1
no, 1, 84
Scientism, 77, 87, 106, 125
Secret of secrets, 16
Selection bias, 17
Self, XV
egoic (limited), XV, 119
deeper, universal (unlimited), XV, 119, 127
Semantic view of theories, 45
Sense writing, 40
Sensing, 39
just, 38, 39-40, 87
Sensory perception, 38
Separation, 126, 127
ending, 131
illusion of, 126

myth of, 126
story, 131
Sexual behaviour, 120
Shadow work, 106
Sheldrake, Rupert, 11, 17, 18, 48, 50-52, 99, 125
Shepherd, Linda Jean, 53-54
Shiva, Vandana, 63
Shultz, J. W., 20
Siegfried, Tom, 8
Silence, 79, 87, 93, 105, 128
Simmel, Georg, 20
Simplicity, 74
Simplification, 42
Society, 59-72,
20 imbalances in, 107-108
Sociobiology, 54
Somers, Suzanne, 98
Sound, 104
healing, 104
SPAC, 90
Space and time, 7-8
Speech, 101-104
Speech (or conference) of Birds, 82
Spielmann, Roger, 36
Spirit, 77
Spiritual experience, 7, 9, 11, 87
Spirituality, 77-83, 87
Spiro, Rupert, 53
State of consciousness, 15
St. Denis, Ruth, 80
Steiner, Rudolf, 15, 56, 97
Stockdale, Steve, 31, 101
Stone, Douglas, 104
Structural Differential, 31, 128
Subjectivity, 15, 19, 20, 80, 84, 87
inter-subjectivity, 15
taboo of, 70
Subtle and causal bodies, 50
Subtle and causal energies, 50
Sunyata, 78
Swans, 5
black, 5

T
Tanahashi, Kazuaki, 78
Tannen, Deborah, 104

Tantra, 120
Taoism or Daoism, 19, 22-23, 54, 122
Tao Te Ching or Daodejing, 37, 80, 81, 88, 116, 121, 123, 129
Tart, Charles, 16
Taylor, Steve, 56
Techno-humanism, 126
Technology, 1, 132
Teicholz, Nina, 17
Tennyson, Alfred, 95
Tesla, Nicola, 58
Testability, 40-41
Theoria (Plato), 40
Theory of colour, 7
Theory of Forms, 122
Theory-ladenness of facts, 2
Thich Nhat Hanh, 78, 81, 83, 127, 128, 131
Thinking, 21-27, 105-106
both/and thinking, 115
fuzzy, 26-27
harmful thinking, 33
healing, 27
healthy, 33
philosophical, 120-123
primeval, 120
scientific, 123-124
skills, 105-106
Yin-Yang, 22-23
Third-person, 8, 78-79
Thought, 118
lost in, 118
Tiller, William A., 50, 97
Tolle, Eckhart, 118, 128, 130
Transcendence, 126-131
Tribes, 111
Tribal mentality, 111
Trikaya, 98
Trinities, 74
Trungpa, Chögyam, 122
Truth, XVI, 13, 16, 19, 20, 37, 38, 86
Tudge, Colin, 49

U
Uncertainty, 4-7, 92
beyond, 7, 8
Undefined terms, 92

Understanding, X, XIV, 84, 129
Unity, 53, 56, 67, 81, 88, 105, 108, 115, 116, 117, 121, 131
  consciousness, 67
Unnamable, III, 24, 29-37, 46, 49, 53, 56-58, 77, 79, 80, 82, 83, 87, 88, 95, 103, 105, 116-117, 122-124, 132
Uniqueness, 12, 87

V
Values and science, 65
Value-ladenness of facts, 2
Varela, Francisco J., XI
Verb-based languages, 36
Vigyan Bhairav Tantra, 11
Virus mania, 59
Vita, Enza, XV,

W
War metaphor in science, 7
Wagner, Andreas, 65
Wahl, Daniel Christoph, 7, 7
Walach, Harald, 52, 67
Wallace, B. Alan, 19, 70
Waller, John, 18
Wangyal, Tenzin Rinpoche, 100-101
Warzel, Charlie, 126
Watts, Alan, 8, 36
Weil, Andrew, 51, 96, 98, 100
White, Lynn, A3
Whitaker, Julian, A2
Whitehead, Alfred North, 12, 36
Wholeness, III, XIV, 19, 29, 35, 50, 66, 68, 87, 90, 96, 97, 103, 108
  undivided, 35, 90, 94
Wholes, 49-50
Whorf, Benjamin, 36
Wilber, Ken, 50, 54-56, 77, 81, 93, 98, 106, 125
Wildwood, Rob, 114
Wilhelm, Richard, 16
Wilson, E. O., 54
Wisdom, 19, 24, 25, 50, 63, 66, 68, 84-85, 92, 101, 108, 112, 120, 122-124, 129
  many-sided, 24-25
Wiseman, Richard, 16
Witnessing mind, 106
  beyond, 107
Wittgenstein, Ludwig, 105,
Wolfe, George, 3, 9
Woodger, J. H. 91
Words, 29-36, 113-114
World Health Organization (WHO), 18, 59, 61, 63, 71-72

Y
Yin/Yang, 22, 94
  symbol, 22
  thinking, 22-23
Young, Shinzen, XIV, 55, 104, 130, 132

Z
Zadeh, Lofti, 26
Zeilinger, Anton, 8
Zen Buddhism, 79
Zimmermann, Walter, 91

CPSIA information can be obtained
at www.ICGtesting.com
Printed in the USA
BVHW052148260621
610536BV00004B/15